Fundamentals
of Nonlinear Optics
of Atomic Gases

Fundamentals
of Nonlinear Optics
# of Atomic Gases

NIKOLAǏ B. DELONE
Professor of Physics, General Physics Institute,
Academy of Sciences of the USSR, Moscow

and

VLADIMIR P. KRAǏNOV
Professor of Physics, ZIL Autoworks Institute, Moscow, USSR

Translated from the Russian by
Dr. Eugene Yankovsky

WILEY

John Wiley & Sons
New York / Chichester / Brisbane / Toronto / Singapore

Library of Congress Cataloging in Publication Data:

Delone, N. B.
    Fundamentals of nonlinear optics of atomic gases.
    (Wiley series in pure and applied optics)
    Translation of: Osnovy nelineĭnoĭ optiki atomarnykh gazov.

    Bibliography: p.
    Includes index.
    1. Nonlinear optics. 2. Gases, Ionized gases—
Optical properties. 3. Laser beams. I. Kraĭnov,
V. P. (Vladimir Pavlovich), 1938– . II. Title.
III. Series.

QC446.2.D4513 1987   535'.2   87-14775
ISBN 0-471-89391-9

Printed in the United States of America

10 9 8 7 6 5 4 3 2 1

Preface

Nonlinear optics came into being at the same time as the laser, about a quarter of a century ago. Since then it has become a well-developed subject of physics with many and varied applications, for it plays a definitive role in many phenomena which result when laser radiation interacts with matter.

There are many monographs devoted directly or indirectly to nonlinear optics and many of these are cited in the introduction to this book. But until now there has been no simple and consistent account of the basics of nonlinear optics that contains a description of all the interrelated phenomena, from the microscopic to the macroscopic. This, then, is the reason for this book. In it, using a simple model of a time-independent interaction of monochromatic light with an atomic gas, we describe the elementary nonlinear processes that emerge for an isolated atom, the optical characteristics of the medium averaged over a large number of atoms and depending on the intensity of the light, and the basic nonlinear optics phenomena observed in the propagation of an intense light wave through the medium.

We have focused on an analytical-theoretical description of nonlinear optics phenomena and we have illustrated conclusions with experimental results. Such an approach reflects the essence science, which is a theoretical interpretation of experimental facts and which subsequently allows one to predict unknown properties of matter. Our objective is not to describe all known nonlinear optics phenomena even within the scope of this simple model. We have, however, tried to single out the main phenomena that qualitatively distinguish nonlinear optics from the common linear optics of weak light fluxes. Although the simple model considered here ignores many aspects of nonlinear optics, the basics are presented quite fully.

In referring to the scientific literature on the subject we have given preference to monographs over reviews and reviews over original work with a view to making it easier for the reader to familiarize himself with additional material.

We have endeavored to present the material in such a way so that it is accessible to a wide range of physicists and engineers as well as to senior/graduate college students. We have assumed that the reader is familiar with the basics of atomic physics, quantum mechanics, the quantum theory of radiation, and the physics of lasers. Readers who require additional information on these subjects should turn to the *Course of Theoretical Physics* by L. D. Landau and E. M. Lifshitz (Volume 2, *The Classical Theory of Fields*, 1975, and Volume 3, *Quantum Mechanics: Nonrelativistic Theory*, 1977), to *Atomic Spectra and Radiative Transitions* by I. I. Sobelman, and to *Principles of Lasers* by O. Zvelto.

The authors are grateful to V. A. Kovarskiĭ for reviewing the manuscript and making many valuable suggestions.

NIKOLAĬ B. DELONE
VLADIMIR P. KRAĬNOV

Moscow
May 1986

Contents

Notation List

LATIN SYMBOLS

a	transverse dimension of light beam
c	speed of light
$\mathbf{d}$	dipole moment
$\mathbf{E}$	electric field strength
$\mathscr{E}_n$	energy in state n
$\mathbf{e}$	polarization vector of light
$\mathbf{E}_{at}$	atomic electric field strength
g	spectral-line factor
$\mathbf{H}$	magnetic field strength
I_ω	intensity of radiation of frequency ω
K	number of photons in a multiphoton process
$\mathbf{k}$	wave vector
l	orbital quantum number
m	magnetic quantum number
M_e	electron mass
n	principal quantum number
n_ω	refractive index at frequency ω
N_n	number of atoms in state n per unit volume
N_ω	number of photons of frequency ω
$\mathbf{P}$	dipole polarization vector of an atom
p	electron momentum
R	reflection coefficient of an electromagnetic wave
S	photon flux density
T, t	time of action of laser pulse
T	temperature of gas
U	atomic potential

V	interaction of radiation with atom
w	transition rate
z_{nm}	dipole matrix element

GREEK SYMBOLS

α	dipole susceptibility
Γ_n	width of state n
γ	adiabaticity parameter
Δ	resonance misfit (detuning)
$\Delta\omega$	spectral linewidth of the radiation
$\delta\mathscr{E}_n$	energy shift of state n
λ	wavelength
ν	frequency of spontaneous or stimulated radiation
ρ	density of energy levels
σ	absorption cross section
τ	natural lifetime of an atomic state
φ	phase of an elliptically polarized field
θ	angle characterizing the ellipticity of a field
$\chi^{(1)}$	linear atomic susceptibility
$\chi^{(1)}_{nn}$	linear susceptibility of an atom in state n
$\chi^{(1)}_{np}$	off-diagonal linear susceptibility
$\chi^{(K)}$	nonlinear susceptibility of order K
Ψ_n	wave function of state n
ω	frequency of incident electromagnetic wave
ω_{mn}	atomic transition frequency
$d\Omega$	solid-angle differential
Ω	Rabi frequency

Fundamentals
of Nonlinear Optics
# of Atomic Gases

Introduction

Optics is the branch of physics that studies light,—that is, the electromagnetic field in the visible range of frequencies—and its interaction with matter. The term *optics* was defined long ago and is fairly unambiguous. The term *nonlinear optics* is more ambiguous. This is quite in the nature of things, since nonlinear optics, as an independent chapter of optics, appeared rather more recently. The birth of nonlinear optics, as is well known occurred at the same time as that of the laser, in the early 1960s. Now by nonlinear optics one usually means that department of optics that studies the interaction of matter with light of high intensity, which changes the optical properties of the matter itself. If, as is customary in optics, we describe the properties of matter by the optical characteristics averaged over a large number of atoms (that is, by the polarization, dielectric constant, and index of refraction), then for light of high intensity all these characteristics become dependent on the electric field strength $\mathbf{E}$ in the light wave. For instance, the polarization vector $\mathbf{P}$ must generally be represented in the form of a series in which the terms are proportional to different powers of the electric field strength:

$$\mathbf{P} = \mathbf{P}^{(1)} + \mathbf{P}^{(2)} + \mathbf{P}^{(3)} + \cdots,$$

with $\mathbf{P}^{(1)} \propto \mathbf{E}$, $\mathbf{P}^{(2)} \propto \mathbf{E}^2$, $\mathbf{P}^{(3)} \propto \mathbf{E}^3$, etc., in the absence of resonances. Here, as everywhere in what follows, we have assumed the use of the atomic system of units ($\hbar = e = M_e = 1$), in terms of which the electric field strength E is always small compared to unity in the real situations we are interested in. Media for which such series expansions are valid and require terms beyond the first are called *nonlinear*. Allowing for the higher-order terms in the expansion of the polarization in powers of the electric field strength is equivalent to allowing for nonlinear scattering and multiphoton light absorption by the medium. For low-intensity light, that is, low electric field strength in the light field, we can ignore the higher-order terms in the expansions in comparison with the first term, which corresponds to the initial principles of ordinary optics. In accordance with this it is logical to call normal optics *linear*. This term, however, has not yet acquired wide application, although it well reflects the essence of the situation. We will be using it.

What is important is that, from the standpoint of the macroscopic phenomena that emerge in the interaction of light with matter, the difference between linear and nonlinear optics is qualitative. For instance, in nonlinear optics the basic laws of the common linear optics, such as the law of linear superposition of light fluxes and the law of rectilinear propagation of light, cease to be valid, and light of frequency ω can excite light of frequencies 2ω, 3ω, etc.

Since nonlinear optics is based on the phenomena associated with the change in the properties of matter under the action of light, it is obvious that a consistent exposition of nonlinear optics cannot be confined to a discussion of macroscopic phenomena; a microscopic discussion of the phenomena at the atomic level is also necessary. A complete and consistent exposition of nonlinear optics requires considering the elementary nonlinear effects that emerge at the atomic level and lead to the dependence of the averaged optical characteristics of matter on the intensity of light; it further requires a description of the macroscopic phenomena that appear during the propagation of a light wave in a medium, by solving the Maxwell equations and allowing for the nonlinear nature of the medium. This book is devoted to just such a thorough and consistent exposition of the basics of nonlinear optics.

One must bear in mind that the microscopic description of the interaction of light with matter was traditionally employed within the scope of the common, linear optics in such fields as molecular optics (see Born and Wolf, 1975; Fabelinskiĭ, 1968; and Vol'kenstein, 1951). Lately several monographs have been published devoted to a microscopic description of the nonlinear-optics phenomena that emerge at the atomic level (see Delone and Kraĭnov, 1985; Kovarskiĭ, 1974; and Rapoport, Zon, and Manakov, 1978). However, in classical monographs devoted to nonlinear optics—for instance, Akhmanov and Khokhlov, 1964, and Bloembergen, 1965—the focus is on the macroscopic approach and solution of the Maxwell equations for nonlinear media. Other monographs can also be cited (Apanasevich, 1977; Butylkin, Kaplan, Khronopulo, and Yakubovich, 1984; Hanna, Yuratich, and Cotter, 1979; Kielich, 1981; and Schubert and Wilhelmi, 1971) that to one degree or another consider the whole range of questions from the microscopic to the macroscopic. But the tendency of the authors of these books to describe in great detail the greatest possible number of nonlinear-optics phenomena in diverse media makes it difficult to determine the main laws of nonlinear optics.

In this book we have tried something different. We have simplified the exposition as much as possible and used a unifed language and the same approximations.

The general approach to describing the interaction is to use the common semiclassical approximation, namely, the atom is described by using quantum mechanics, while the electromagnetic field is described in terms of classical field theory. The quantum term "photon" is used only to describe the graphic transition schemes qualitatively and to clarify the meaning of energy conservation in a variable field. The classical terms "intensity of light" and "field strength in the light wave" are used equivalently, depending on the specific problem. For well-known reasons the interaction of light with an atom is described in the dipole approximation. As for the atomic transitions initiated by the light field, we assume that only one optical electron participates in such transitions.

Depending on the phenomenon, in describing the interaction of light with matter, we will be using both the language of wave optics, based on the Maxwell equations, and the language of geometrical optics, which is the classical limit of wave optics and is applicable when the dimensions of the sample are large compared with the wavelength of the light.

For the sake of simplicity we restrict our discussions to the simplest possible model of the interaction; within the scope of this model the light is assumed to be ideally monochromatic and the medium is a gas consisting of isolated atoms. It should be noted that this model is fully realistic though seemingly simple. Indeed, although laser radiation is actually quasimonochromatic, its degree of monochromaticity is very high (up to $\Delta\omega/\omega \sim 10^{-8}$). Furthermore, over a broad range of pressures, a gas can be considered as a medium consisting of isolated atoms. Finally, if we remain strictly within the optical frequency range, the differences between atomic and molecular gases are insignificant because of the small differences in the electronic spectra of atoms and molecules. Of course, in using such a model, the nonlinear-optics phenomena that occur in condensed transparent media (liquids, crystals, and glasses) are excluded in principle. However, notwithstanding the marked differences in the structure of condensed media (the presence of free electrons, the crystal lattice, anisotropy, etc.) and rarefied atomic gases, the main features of nonlinear-optics phenomena are qualitatively the same in all media. The differences are only quantitative.

Another aspect of our approach is that we deal only with steady-state processes. Processes are called *steady-state* when light acts on a substance for a much longer period than the characteristic relaxation times of the substance. When this time interval is shorter than the relaxation time, we will speak of *non-steady-state* (or *transient*) interaction. Quite naturally, such transient nonlinear-optics effects as self-induced transparency, light echo, and optical nutation (e.g. see Apanasevich, 1977) remain outside the scope of this discussion. All these effects differ qualitatively from steady-state effects. It must be noted, however, that the physical essence of these effects is related not to the nonlinearity of the medium but to the ultrashort duration of the exciting light pulse. For this reason, limiting ourselves to the time-independent interaction of light with matter still allows us to describe the basic steady-state nonlinear-optics phenomena.

In most cases we restrict our discussion to the *unperturbed-field approximation* for the light wave incident on the medium. This means that in calculating the nonlinear polarization of the medium, which enters into the macroscopic Maxwell equations in the presence of a medium, we substitute not the true electromagnetic field in the medium, which itself must be found by solving these equations, but the known field in the incident electromagnetic wave. For the unperturbed-field approximation to be valid, it is necessary, first, that the gaseous medium be sufficiently rarefied; second, that the electric field strength in the incident electromagnetic wave be low compared with the characteristic intraatomic fields; and third, that the dimensions of the gaseous medium be not too great. These conditions become more rigorous when resonances develop between the frequency of the field and that of the atomic transitions. This book studies the quantitative conditions for applying the unperturbed-field approximation. Such a restriction, on the one hand, essentially simplifies the mathematical description of nonlinear-optics phenomena, and on the other, does not lead to the loss of any qualitative features of the interaction. Allowing for a substantial change in the incident wave in the medium leads only to quantitative differences from the unperturbed-field approximation.

Finally, we have not attempted to describe all possible variants of the nonlinear interaction in each specific case. The purpose has been the opposite: to set out the basic physical meaning of a given phenomenon and cite the most important examples of its manifestation during the interaction of intense light with an atomic gas.

This book widely uses the language of Feynman diagrams for the linear and nonlinear susceptibilities of an atom, a language taken from quantum electrodynamics. This language makes it possible to classify in a simple and graphic form the different processes in the interaction of light with an atom, proceeding only from the causality principle and the laws of conservation in the absorption and emission of photons by an atom.

The terms "coherent processes" and "incoherent processes" are often used in this book. What is the meaning of these concepts?

First, coherent and incoherent processes differ in the nature of the atomic transitions that take place in the field of an electromagnetic wave. If during an atomic transition a photon of the incident electromagnetic wave is absorbed and a spontaneous photon is emitted that has a different direction of propagation or a different polarization or a different frequency than the incident photons (e.g., in spontaneous Raman scattering), such a transition is said to be *incoherent*. In a *coherent transition* (induced emission) the emitted photon has the same polarization, direction of propagation, and frequency as the photons in the medium (e.g., in stimulated Raman scattering).

Hence, the excitation and relaxation processes are independent in the incoherent case and are interdependent in the coherent case. The relation between excitation and relaxation depends on the population of the excited atomic level. If the level is populated, the atom remains for a long time in this state and "forgets" the population mechanism: excitation is not connected with relaxation, and the entire process is incoherent. The atomic energy levels figure in the corresponding conservation laws together with the photon energies. But if the atomic level is virtual, that is, not populated, the atom lives a short time in this state (this time is determined by the Heisenberg energy-time uncertainty relation). Relaxation in this case is closely connected with excitation, and the entire process is coherent. The energies of the photons, not those of the atomic levels, participate in the corresponding conservation law.

Second, we use the terms "coherent radiation" and "incoherent radiation." By *coherent radiation* we mean a plane monochromatic wave, coherent in both space and time. The radiation is *incoherent* if it has no definite direction of propagation in space, that is, no fixed direction or fixed value of the wave vector, or if the spread in the frequencies of radiation or in the phases of separate monochromatic components is great.

Third, in the macroscopic nonlinear-optics processes that emerge in the interaction of light with matter, we refer to the interaction as coherent if the so-called phase-matching condition is required for this interaction to have an appreciable probability (Section 3.1). This condition means that as the exciting wave of frequency ω_1 propagates in the medium, the interrelationship between the phase of the exciting wave and that of the excited wave (frequency ω_2) remains unchanged. The interaction is called incoherent if this condition is not required. We will see in Section 3.1 that for coherent processes (in the above-noted sense) the probabilities of optical processes are proportional to the square of the number of atoms in the medium, while for incoherent processes the probabilities are proportional to the first power. This difference explains the origin of these names: in coherent processes the *amplitudes* of the processes at different atoms of the medium are added, while in incoherent ones their *probabilities are added*.

Finally, a few words must be said to justify Chapter 1, devoted to linear optics. Although the basic laws of linear optics are well known and have been detailed in many textbooks (e.g. see Born and Wolf, 1975, and Ditchburn, 1963), the principal reason we have included this chapter is to be able to compare nonlinear and linear optics in one book. For this reason the linear optics in Chapter 1 is outlined within the scope of the same model and in the same language as the nonlinear optics of Chapters 2 and 3. The exposition of linear optics in Chapter 1 does not duplicate the exposition in the above-cited literature.

It is well known that nonlinear optics has many applications. We may mention nonlinear spectroscopy and new sources and generators of coherent radiation. This book does not undertake to examine these applications. The interested reader can find many monographs on the subject (Akhmanov and Koroteev, 1981; Bloembergen, 1977; Dmitriev and Tarasov, 1982; Letokhov, 1982; Letokhov and Chebotaev, 1977; Rautian, Smirnov, and Shalagin, 1979; Walther, 1976; and Zernike and Midwinter, 1973).

1

Linear Optics of Atomic Gases

The linear optics of atomic gases, that is, the theory of the interaction of a weak electromagnetic field of optical frequency with a gas, is based on the linear effects of the interaction of this field with an isolated atom of the gas considered. The term linear means that in the process of this interaction the atom absorbs only one photon of the external electromagnetic field and goes from the ground state to an excited state.

Generally speaking, optics is a wave theory, that is, it describes the interaction of an electromagnetic field with a medium. But if the dimensions of the medium are large compared with the wavelength of the electromagnetic field of the radiation, then the limit of geometrical optics is valid. The basic laws of geometrical optics are well known (e.g., see Ditchburn, 1963). These are the rectilinear propagation of light in a homogeneous gaseous medium, the superposition of light beams (which follows from the linearity of Maxwell's equations in a medium), the equality of the angle of incidence and the angle of reflection from a plane boundary between vacuum and a gaseous medium, and Snell's law for the refraction of light in a medium (which relates the angle of incidence of a light beam on a medium and the angle of the beam refracted by the medium).

For the interaction of light with an atom, we will restrict our discussion, as has already been said in the Introduction, to the dipole approximation, which means that the operator of the interaction is

$$V = \mathbf{r} \cdot \mathbf{E}(t), \qquad (1.1)$$

where $\mathbf{E}(t)$ is the electric field strength in the electromagnetic wave, and $\mathbf{r}$ is the position vector of the atomic electron undergoing an optical transition caused by the light field. The dipole approximation is applicable when the atom's dimensions are small compared with the wavelength of the light radiation. Equation (1.1) is written in the atomic system of units; the majority of mathematical expressions in this book are given in this system, as noted in the Introduction. Note that the Planck constant, the electron charge, and the electron mass are equal to unity in this system.

Linear optics of gaseous media, as was noted above, is based on the assumption that the interaction of the light field with the atom is linear. This is true for low fields and also for rarefied gases (the numerical estimates are given below). Due to the small density of the gas, the effects of separate atoms are additive, so that to obtain macroscopic characteristics, say per unit volume of the gas, we need only multiply the corresponding characteristic of a single atom by the number N of atoms per unit volume.

A special feature of the exposition in this chapter is that we employ the unperturbed electromagnetic field approximation when solving Maxwell's equations for the propagation of electromagnetic waves in linear optics. Of course, this is far from necessary, since the linear Maxwell equations are easily solved without the above restriction. But since there will be a further discussion of the nonlinear-optics effects in Chapter 3, and these are often impossible to analyze without resorting to this approximation, we will employ this approximation in our discussion of linear optics for the sake of unity of exposition. In linear optics rejection of this approximation leads to no new results, but only changes the results quantitatively. We will restrict our discussion, just as in nonlinear optics, to steady-state processes.

In this chapter we will not give an exhaustive coverage of the effects present in linear optics. We wish only to touch on the basic phenomena that are the linear analog of the main nonlinear-optics phenomena discussed in Chapters 2 and 3.

1.1. THE NONRESONANCE LINEAR ATOMIC SUSCEPTIBILITY

1.1.1. Introduction

The most important characteristic of the linear interaction of an electromagnetic field with an isolated atom is the linear polarization of the atom in the electric field of the optical radiation. By definition this quantity is the induced dipole moment $\mathbf{r}$ averaged over the quantum-mechanical state of the atom, $\Psi_n(\mathbf{r}, t, \mathbf{E}(t))$. The wave function Ψ_n describes the state of the atom in the field $\mathbf{E}(t)$. Thus, for the polarization vector we arrive at the following formula:

$$\mathbf{P}(t) = \langle \Psi_n | \mathbf{r} | \Psi_n \rangle. \tag{1.2}$$

In a weak electromagnetic field this vector can be expanded in a Taylor series in powers of $\mathbf{E}(t)$. The leading term in this series is

$$P_i^{(1)}(t) = \sum_j \int_0^\infty \chi_{ij}^{(1)}(\tau) E_j(t - \tau) \, d\tau. \tag{1.3}$$

The integral with respect to time in (1.3) reflects the fact that the polarization $\mathbf{P}$ of the atom at time t is determined, in accordance with the causality principle, by the values of $\mathbf{E}$ at all previous times $t - \tau$. The coefficients $\chi_{ij}^{(1)}(\tau)$ in (1.3) form the *linear susceptibility tensor* of the atom. The subscripts i and j stand for the three projections on the coordinate axes.

Let us take the example of a monochromatic field $\mathbf{E}(t) = 2\mathbf{E}_\omega \cos \omega t$. Expanding polarization $\mathbf{P}^{(1)}(t)$ in a Fourier series, that is,

$$\mathbf{P}^{(1)}(t) = \sum_\omega \mathbf{P}_\omega^{(1)} \exp(i\omega t), \tag{1.4}$$

introducing

$$\chi_{ij}^{(1)}(\omega; \omega) = \int_0^\infty \chi_{ij}^{(1)}(\tau) \exp(-i\omega\tau) \, d\tau, \tag{1.5}$$

and employing (1.3), we arrive at the following relationship between $\mathbf{P}_\omega^{(1)}$ and $\mathbf{E}_\omega$:

$$P_{i\omega}^{(1)} = \sum_j \chi_{ij}^{(1)}(\omega; \omega) E_{j\omega}. \tag{1.6}$$

Note that it is the Fourier component $\mathbf{P}_\omega^{(1)}$ that is related to the experimentally observed quantities in the linear interaction of light with a gaseous medium.

In the linear susceptibility $\chi_{ij}^{(1)}(\omega; \omega)$ in (1.6), the first independent variable ω is the frequency at which polarization $\mathbf{P}_\omega^{(1)}$ of the atom occurs, while the second is the frequency of the external electromagnetic field. In the given linear case the two quantities coincide. As we will see in Chapter 2, in the nonlinear case this may not be true.

The expression $\chi_{ij}^{(1)}(\omega; \omega)$ for the linear susceptibility corresponds to the diagonal matrix element (1.2) of the dipole moment operator with respect to the state n of the atom. We will see that this susceptibility describes linear scattering, a process in which the atom absorbs a single photon from the external electromagnetic field and then, emitting a spontaneous photon, returns to its initial state n. But if the linear scattering process results in the optical electron going over to a state p that differs from the initial state, the process obviously is described by the off-diagonal matrix element of the dipole moment operator for the atom with respect to the initial (n) and final (p) states. When necessary, we will attach subscripts referring to the initial and final states to $\chi^{(1)}$, for instance, $\chi_{nn}^{(1)}$ in (1.6). The quantity $\chi_{np}^{(1)}$ is defined in a similar manner, the only difference being that here the frequency ν with which the polarization vector $\mathbf{P}^{(1)}(t)$ oscillates does not necessarily coincide with the frequency ω of the external electromagnetic field $\mathbf{E}(t)$; the relation between the two frequencies is $\nu = \omega - \omega_{pn}$. In this case we are dealing with $\chi_{np}^{(1)}(\nu; \omega)$, which corresponds to spontaneous Raman scattering, with ν the frequency of the emitted spontaneous photon. If $p = n$, we are dealing with Rayleigh scattering of light by an atom. We note once more that both Rayleigh scattering and spontaneous Raman scattering are processes in which only one photon of the external electromagnetic field is absorbed.

The electric field strength in a monochromatic electromagnetic wave depends also on the position vector $\mathbf{r}$:

$$\mathbf{E}_\omega(\mathbf{r}) = \sum_\mathbf{k} \mathbf{E}_\omega(\mathbf{k})\exp(-i\mathbf{k} \cdot \mathbf{r}). \tag{1.7}$$

the vector $\mathbf{P}_\omega^{(1)}$ (k) is defined in a similar manner. In both cases $\mathbf{k}$ is called the *wave vector*. Its length and the frequency ω are related thus: $k = \omega/c$. Since in a homogeneous medium the linear susceptibility $\chi^{(1)}$ is independent of $\mathbf{r}$, substituting (1.7) into (1.6) readily yields

$$P_{i\omega}^{(1)}(\mathbf{k}) = \sum_j \chi_{ij}^{(1)}(\omega; \omega) E_{j\omega}(\mathbf{k}). \tag{1.8}$$

For brevity we will usually drop the $\mathbf{k}$ in $\mathbf{P}_\omega^{(1)}(\mathbf{k})$ and $\mathbf{E}_\omega(\mathbf{k})$.

1.1.2. The Diagrammatic Technique for Linear Atomic Susceptibility

Section 2.3 of Delone and Kraĭnov, 1985, provides a detailed account of the diagrammatic technique needed to describe the interaction of a monochromatic electromagnetic field with an isolated atom. Here we give only a brief exposition of this technique without going into substantiation, which requires the quantum-mechanical theory of time-dependent perturbations for the wave function.

According to this technique, the linear susceptibility $\chi_{ij}^{(1)}(\omega; \omega)$ is the two-photon matrix element of the dipole moment operator $\mathbf{r}$ and is depicted by the two Feynman diagrams shown in Figure 1.1. A dashed line denotes the absorption of a single photon from the external electromagnetic field of frequency ω, while a wavy line denotes the spontaneous emission of a photon, whose frequency ν in this case is also ω. Thus, the two Feynman diagrams are first-order diagrams in the external electromagnetic field.

Circles in the diagram stand for dipole matrix elements, say $\mathbf{r}_{mn}$. A vertical line between two circles corresponds to a propagator in which from the energy of a state denoted by a set of quantum numbers (in our case m) we must subtract the energy of the initial state n and the energy of the

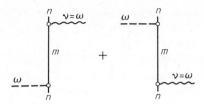

Figure 1.1. Feynman diagrams for linear atomic susceptibility.

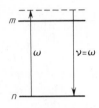

Figure 1.2. Emission-absorption diagram for Rayleigh scattering of light on an atom, corresponding to the first diagram shown in Figure 1.1.

photon absorbed in the transition from n to m (see the first diagram in Figure 1.1). For instance, the denominator in the propagator for the vertical line in the first Feynman diagram in Figure 1.1 (line m) is $\omega_{mn} - \omega$. Finally, the rules for operating with Feynman diagrams require summation (or integration) over the quantum numbers of the intermediate state m of the atom.

According to their form, the Feynman diagrams in Figure 1.1 define the *diagonal* matrix element with respect to state n of the atom. *Off-diagonal* matrix elements, in which the initial and final states do not coincide, can be determined in a similar manner. We touch on these diagrams in Section 1.1.5.

The Feynman diagrams in Figure 1.1 describe the Rayleigh scattering of light on an atom. Figure 1.2 shows the process of absorption and emission of a photon for the first diagram in Figure 1.1 in the case considered here.

1.1.3. Properties of Linear Atomic Susceptibility

In accordance with the rules for evaluating the Feynman diagrams of Figure 1.1, we arrive at the following expression for the linear atomic susceptibility:

$$\chi_{ij}^{(1)}(\omega; \omega) = \sum_m \left[r_{nm}^i r_{mn}^j (\omega_{mn} - \omega)^{-1} + r_{mn}^i r_{nm}^j (\omega_{mn} + \omega)^{-1} \right]. \quad (1.9)$$

We can easily see that because of the conservation of parity of atomic states the quantity $\chi_{ij}^{(1)}$ is diagonal in i, j, or

$$\chi_{ij}^{(1)} = \chi^{(1)} \delta_{ij}. \quad (1.10)$$

Indeed, let us take, for the sake of definiteness, an electromagnetic field that is linearly polarized along the z axis. Then $j = z$ in (1.9). Since the magnetic quantum numbers of states n and m are the same, due to the selection rules for the magnetic quantum number in the dipole matrix element z_{mn}, the magnitude r^i_{nm} in (1.9) is nonzero only if $i = z$. But if $i = x$ or $i = y$, these operators, as is known from quantum mechanics, change the value of the magnetic quantum number in a transition. This proves the validity of (1.10) for the case of a linearly polarized field. A similar proof can be carried out for an electromagnetic field with arbitrary polarization. From the classical viewpoint, Eq. (1.10) follows from the equal probability for all the orientations of the atom in space.

Equation (1.9) shows that the term with $m = n$ is absent from the sum over the virtual states m (in this connection see Section 2.1). Therefore, when we go over to the static limit $\omega \to 0$, this equation leads to the static dipole susceptibility of the atom in state n:

$$\chi^{(1)}(0;0) = 2 \sum_{m \neq n} |z_{mn}|^2 \omega_{mn}^{-1}. \tag{1.11}$$

This corresponds to the well-known second-order term in the quantum-mechanical perturbation expansion in powers of the interaction (1.1).

In the other limit of high frequencies, $\omega \gg \omega_{mn}$, combining (1.9) with the dipole sum rule (see Bethe and Salpeter, 1957) easily yields

$$\chi^{(1)}(\omega; \omega) = -\omega^{-2}. \tag{1.12}$$

This formula is valid for a linearly polarized field.

1.1.4. The Quadratic AC Stark Effect in a Monochromatic Field

If we ignore the hydrogen-atom levels and the hydrogenlike highly excited (Rydberg) levels of complex atoms, then, as is well known, the atomic levels in a constant electric field experience a quadratic Stark shift (see Bethe and Salpeter, 1957). In a monochromatic varying field $\mathbf{E}(t) = 2\mathbf{E}_\omega \cos \omega t$ the levels, under certain conditions (see below), experience a quadratic ac Stark shift, whose value is (see Delone and Kraĭnov, 1985, Section 6.2)

$$\delta \mathscr{E}_n = -\chi^{(1)}(\omega; \omega) E_\omega^2$$
$$= -E_\omega^2 \sum_m |z_{mn}|^2 \left[(\omega_{mn} - \omega)^{-1} + (\omega_{mn} + \omega)^{-1} \right]. \tag{1.13}$$

This expression is valid for the case of an electromagnetic field linearly polarized along the z axis. Thus, the shift is determined by the linear susceptibility $\chi^{(1)}(\omega; \omega)$, with the associated Feynman diagrams shown in Figure 1.1.

In the static limit, $\omega \to 0$, combining (1.13) with (1.11) leads to the well-known dc limit for the Stark shift in a constant field:

$$\delta\mathscr{E}_n = -2E^2 \sum_{m=n} |z_{mn}|^2 \omega_{mn}^{-1}. \tag{1.14}$$

In the opposite case, $\omega \gg \omega_{mn}$, combining (1.13) with (1.12) yields an expression that corresponds to classical oscillations of an optical electron in the field of an electromagnetic wave, with the effect of the atomic field being so small that it can be ignored. The result is

$$\delta\mathscr{E}_n = E_\omega^2 \omega^{-2}. \tag{1.15}$$

If there are several valence electrons in the unfilled shell, we must multiply (1.15) by the number of such electrons, since the free oscillations of different electrons do not interfere with each other.

For the quadratic ac Stark shift (1.13) to manifest itself (see Delone and Kraĭnov, 1985, Chapter 6), it must be small compared to the energy $\hbar\omega_{mn}$ of characteristic atomic transitions or the energy $\hbar\omega$ of a photon from the electromagnetic field:

$$\delta\mathscr{E}_n \ll \hbar\omega_{mn} \quad \text{or} \quad \delta\mathscr{E}_n \ll \hbar\omega. \tag{1.16}$$

1.1.5. Linear Scattering of Light

As we have already seen, in an external electromagnetic field an atom acquires an induced dipole moment $\mathbf{P}^{(1)}$ that oscillates periodically with a frequency ω. As is known from the semiclassical theory of radiation (see Landau and Lifshitz, 1975, §66), the atom will emit photons at the same frequency.

The probability rate for spontaneous emission of a photon of frequency $\nu = \omega$ by an oscillating dipole moment $\mathbf{P}^{(1)}$ is given by the well-known formula (Berestetskiĭ, Lifshitz, and Pitaevskiĭ, 1980, §45)

$$dw_{nn} = \frac{\omega^3}{2\pi c^3} \left| \mathbf{e} \cdot \mathbf{P}_\omega^{(1)} \right|^2 d\Omega, \tag{1.17}$$

where $\mathbf{e}$ is the unit vector specifying the polarization of the emitted spontaneous photon, and $d\Omega$ is the solid angle within which the photon is emitted. Assuming the external field to be linearly polarized along the z axis and substituting the expression (1.6) for $\mathbf{P}_\omega^{(1)}$ into (1.17), we can relate the probability of spontaneous emission of a photon to the linear susceptibility of the atom through the following formula:

$$dw_{nn} = \frac{\omega^3}{2\pi c^3} \left| \chi^{(1)}(\omega; \omega) \right|^2 e_z^2 E_\omega^2 \, d\Omega. \tag{1.18}$$

Here, in accordance with (1.13), we have

$$\chi^{(1)}(\omega; \omega) = \sum_m |z_{mn}|^2 \left[(\omega_{mn} - \omega)^{-1} + (\omega_{mn} + \omega)^{-1} \right]. \tag{1.19}$$

Equation (1.18) gives the probability of Rayleigh (elastic) scattering of light on an atom, since the emitted photon has the same frequency ω as the photon from the incident electromagnetic field. However, the directions of the propagation of the two photons do not necessarily coincide. Dividing (1.18) by the flux density of the incident photons, we can find the Rayleigh scattering cross section. Here we will not give the appropriate formulas (Heitler, 1954), since our main aim is to establish the relationship between the linear susceptibility of an atom and the linear scattering of light by the atom.

We have so far assumed that the final state of the atom coincides with the initial one. If they differ, we are dealing with spontaneous Raman scattering. After absorbing a photon from an external electromagnetic field of frequency ω, an atom that was in initial state n goes over to a virtual intermediate state m. It then emits a spontaneous photon of frequency ν and goes over to its final state p, so that $\nu = \omega - \omega_{pn}$. If we assume that n is the ground state of the atom, then, obviously, ν is always lower than ω (the schematic diagram for spontaneous Raman scattering is shown in Figure 1.3a). The frequency ν, in this case, is known as the *Stokes frequency*. If n is an excited state, ν may be higher than ω, in which case ν is known as the *anti-Stokes frequency* (Figure 1.3b).

The probability rate for spontaneous Raman scattering is determined in a manner similar to (1.16):

$$dw_{np} = \frac{\nu^3}{2\pi c^3} \left| \chi_{np}^{(1)}(\nu; \omega) \right|^2 E_\omega^2 \, d\Omega. \tag{1.20}$$

The susceptibility $\chi_{np}^{(1)}(\nu; \omega)$ is determined by the Feynman diagrams shown in Figure 1.4 and is given by the following formula:

$$\chi_{np}^{(1)}(\nu; \omega) = \sum_m \left\{ \frac{(\mathbf{r}_{pm} \cdot \mathbf{e})(\mathbf{r}_{mn} \cdot \mathbf{e}')}{\omega_{mn} - \omega} + \frac{(\mathbf{r}_{pm} \cdot \mathbf{e}')(\mathbf{r}_{mn} \cdot \mathbf{e})}{\omega_{mn} + \omega} \right\}. \tag{1.21}$$

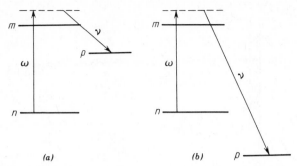

Figure 1.3. Emission-absorption diagram for spontaneous Raman scattering: (*a*) Stokes scattering, and (*b*) anti-Stokes scattering.

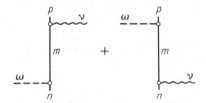

Figure 1.4. Feynman diagrams for the off-diagonal matrix element of the linear atomic susceptibility describing spontaneous Raman scattering.

Here **e** is the unit polarization vector for the emitted spontaneous photon with a frequency ν, and $\mathbf{e}' = \mathbf{E}_\omega/E_\omega$ is the unit polarization vector of an incident photon from the external electromagnetic field with a frequency ω.

In contrast to Rayleigh scattering, in the case of spontaneous Raman scattering the spontaneously emitted photon may be polarized along the x or the y axis.

1.1.6. Calculation of the Linear Atomic Susceptibility

As Eq. (1.19) or (1.21) shows, there are two difficulties in calculating the linear atomic susceptibility. The first lies in the fact that one has to sum over an infinite number of intermediate atomic states m, including the states from the continuous spectrum. In some cases this difficulty can be overcome by employing either the atomic Green function (Delone and Kraĭnov, 1985, Section 2.6) or commutation relations (Landau and Lifshitz, 1977, p. 286, Problem 4). It must be noted at this point that the often employed direct summation in (1.19) over states m that lie close to state n usually results in large errors because the contribution of the continuous spectrum is ignored.

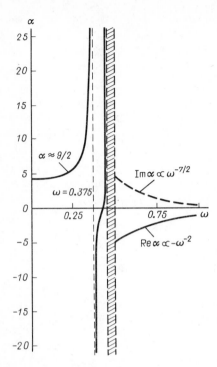

Figure 1.5. Linear susceptibility of the ground state of a hydrogen atom as a function of the frequency ω of a linearly polarized external electromagnetic field. The hatched section corresponds to the region near the boundary of the continuous spectrum with a large number of closely spaced resonances. At $\omega > 0.5$ the susceptibility is complex-valued: its real part is shown by a solid curve, and its imaginary part by a dashed curve. All quantities are given in atomic units.

The second difficulty is the absence of exact expressions for the wave functions of the atom unperturbed by the field. Here one usually uses the quantum-defect approximation, the model-potential method (Delone and Kraĭnov, 1985, Section 2.6) and the Hartree–Fock approximation (Landau and Lifshitz, 1977, §69). Within the scope of the first two methods it has proved possible to build an analytical expression for the atomic Green function, although the actual calculation of the linear susceptibility using the Green function requires a computer because of the complexity of the special functions in whose terms the Green function is expressed.

The calculation of the linear susceptibility $\alpha \equiv \chi^{(1)}$ for the ground state of the hydrogen atom is the simplest. This was done by Gavrila, 1967, who used the Coulomb Green function. The results are depicted in Figure 1.5. One can easily see that the linear susceptibility expressed as a function of frequency ω has resonances, in accordance with (1.19), at frequencies $\omega = \omega_{mn}$.

Figure 1.6 shows $\alpha = \chi^{(1)}(\omega; \omega)$ as a function of ω for the ground state of a xenon atom (Davydkin, Zon, Manakov, and Rapoport, 1971). As in Figure 1.5, here we have one-photon resonances, with $\chi^{(1)}(\omega; \omega)$ vanishing between the resonances.

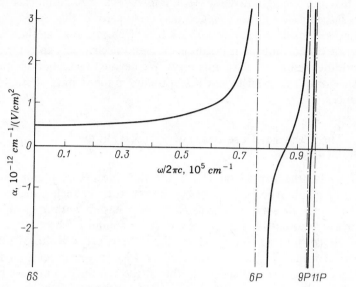

Figure 1.6. Linear susceptibility of ground state of a xenon atom as a function of frequency ω of external electromagnetic field (the field is linearly polarized).

1.1.7. Conclusion

To summarize, we can say that the linear atomic susceptibility is the basic characteristic of the interaction of a weak electromagnetic field with an atom, since in terms of this quantity we can express both the quadratic ac Stark shift of the energy levels of the atom in a monochromatic field and the probabilities for processes of linear scattering of light by the atom, namely, Rayleigh scattering and spontaneous Raman scattering. We also note that the linearity of the interaction of light with an atom is equivalent to the one-photon nature of atomic transitions in a light field.

In the next section we will see that such processes as one-photon ionization and linear absorption are also closely linked with the linear atomic susceptibility.

1.2. THE RESONANCE LINEAR ATOMIC SUSCEPTIBILITY

1.2.1. Introduction

As we have just seen, in the case where $\omega \approx \omega_{mn}$, when the linear susceptibility (1.19) is resonantly high, the deviations from the resonances are so small that any changes in these deviations have a strong influence on the linear susceptibility. The changes may result from the Stark shift of resonances considered in Section 1.1.4 or from various mechanisms of resonance broadening (it is because of these mechanisms that the linear susceptibility remains a finite quantity). We devote this section to studying the behavior of the linear atomic susceptibility in the vicinity of one-photon resonances.

1.2.2. The Frequency Dependence of $\chi^{(1)}(\omega; \omega)$ in the Vicinity of Resonances

Qualitatively, the expression (1.19) for the linear susceptibility in the vicinity of a resonance, $\omega \approx \omega_{mn}$, can be rewritten in such a manner that only the resonance term is isolated from the sum with respect to m. Employing the Breit–Wigner procedure (Landau and Lifshitz, 1977, §134), we introduce the term $-i\Gamma_m$, where Γ_m is the effective width of level m, into the propagator (the denominator). This leads to a Lorentzian-shaped resonance curve. Strictly speaking, this shape is present only for certain broadening mechanisms, such as the broadening caused by spontaneous transitions from the excited state m, but we will use this procedure everywhere, since we are interested primarily in the qualitative picture in the vicinity of resonances.

Equation (1.19) therefore yields

$$\chi^{(1)}(\omega; \omega) = |z_{mn}|^2 (\omega_{mn} - \omega + i\Gamma_m)^{-1}. \tag{1.22}$$

We see that in the vicinity of resonances the linear susceptibility becomes complex-valued.

If we introduce (1.22) into (1.18), we find the probability for linear resonance Rayleigh scattering (resonance fluorescence):

$$dw_{nn} = \frac{\omega^3}{2\pi c^3} \frac{|z_{mn}|^4 e_z^2 E_\omega^2 \, d\Omega}{(\omega_{mn} - \omega)^2 + \Gamma_m^2}. \tag{1.23}$$

If we introduce the real part of (1.22) into (1.13), we find the resonance quadratic ac Stark shift of level n in a monochromatic field:

$$\delta\mathscr{E}_n = -\frac{|z_{mn}|^2 (\omega_{mn} - \omega) E_\omega^2}{(\omega_{mn} - \omega)^2 + \Gamma_m^2}. \tag{1.24}$$

In particular, at an exact resonance, $\omega = \omega_{mn}$, the quantity $\delta\mathscr{E}_n$, according to (1.24), vanishes, instead of becoming infinite as one might suppose from (1.13). The ac Stark shift changes sign at an exact resonance.

Similarly, applying (1.21) to spontaneous Raman scattering and employing the Breit–Wigner procedure, we find that

$$\chi_{np}^{(1)} = (\mathbf{r}_{pm} \cdot \mathbf{e})(\mathbf{r}_{mn} \cdot \mathbf{e}')(\omega_{mn} - \omega + i\Gamma_m)^{-1}. \tag{1.25}$$

Substituting (1.25) into (1.20), we find the probability rate for resonance spontaneous Raman scattering:

$$dw_{np} = \frac{\nu^3}{2\pi c^3} \frac{\left|(\mathbf{r}_{pm} \cdot \mathbf{e})(\mathbf{r}_{mn} \cdot \mathbf{e}')\right|^2 E_\omega^2 \, d\Omega}{(\omega_{mn} - \omega)^2 + \Gamma_m^2}. \tag{1.26}$$

Note that here $\mathbf{e}'$ is the unit polarization vector of the incident electromagnetic field, that is, $\mathbf{e}' = \mathbf{E}_\omega / E_\omega$, and $\mathbf{e}$ is the unit polarization vector of the spontaneous photon emitted in the spontaneous Raman scattering process.

Equation (1.26) gives the total probability rate for emission of spontaneous photons with a frequency $\nu = \omega - \omega_{pn}$. But since state p has a certain width Γ_p, we would like to know the probability rate of the spontaneous Raman scattering for the emission of a spontaneous photon with a frequency ν lying in the vicinity of $\omega - \omega_{pn}$. To this end we must multiply (1.26) by a dimensionless factor $g(\nu)\, d\nu$, which determines the shape of the spectral line of the emitted spontaneous photon. Here

$$g(\nu) = \frac{\Gamma_p}{\pi}\left[(\nu - \omega + \omega_{pn})^2 + \Gamma_p^2\right]^{-1}, \tag{1.27}$$

with

$$\int_{-\infty}^{\infty} g(\nu)\, d\nu = 1. \tag{1.28}$$

1.2.3. One-Photon Ionization of an Atom

If the frequency ω of the external electromagnetic field is higher than the energy $\mathscr{E}_n$ of state n for which the linear susceptibility is being calculated, the resonance condition $\omega_{mn} = \omega$ is always satisfied for a state m in the continuous spectrum. Introducing into (1.19) the infinitesimal width of state m and summing over the states of continuous spectrum lying in the vicinity of m, we find that the imaginary part of the linear susceptibility has the form

$$\operatorname{Im}\chi^{(1)}(\omega; \omega) = \pi|z_{mn}|^2 p_m (2\pi)^{-3}. \tag{1.29}$$

Here p_m is the electron momentum in the final state m of the continuous spectrum, and according to the energy conservation law in one-photon ionization we have $\mathscr{E}_m = p_m^2/2 = \omega + \mathscr{E}_n$. Equation (1.29) can be related to the probability rate for one-photon ionization of state n. According to the Fermi "golden rule" (Delone and Kraĭnov, 1985, Section 2.1),

$$w_n = 2 \operatorname{Im} \chi^{(1)}(\omega; \omega) E_\omega^2. \tag{1.30}$$

This expression relates the probability of one-photon ionization to the imaginary part of the linear susceptibility. Hence, the imaginary part of the linear susceptibility at $\omega > \mathscr{E}_n$ corresponds to radiation absorption accompanied by the formation of electrons and ions. This absorption is a one-photon process and diminishes the intensity of the optical radiation when the latter passes through a gaseous medium (see Section 1.4). However, for the optical range of frequencies of the atoms in their ground states, ω is always less than $\mathscr{E}_n$, which is just the opposite of the case discussed above, so that such an absorption mechanism is absent, as a rule. For atoms in excited states the above formulas are realistic and will be used in Section 1.4.

1.2.4. Phenomena that Determine the Resonance Width

Here we will consider the various physical reasons that lead to nonzero imaginary parts Γ_m in the above formulas for the resonance linear susceptibility. We will again ignore the profile of the resonance contour, which obviously may differ for different cases. Irrespective of the profile, it can be stated that the real width of a resonance is determined by whichever of the abovementioned widths prove to be the greatest. We will also point out situations when a resonance splits into two resonances with different widths in a strong light field (see Section 1.2.5). We assume all along that the resonance state m belongs to the discrete spectrum.

1.2.4.1. Natural Broadening. This broadening is caused by the interaction of excited atomic states with the electromagnetic field associated with the physical vacuum. As a result of this interaction, spontaneous electromagnetic dipole transitions on lower atomic states occur. The transition from state m to state n may be accompanied by emission of a single photon, or several photons in the case of a cascade process of spontaneous deexcitation of the excited state m if between n and m there are other excited atomic states. The natural width of state m, which is equal to the probability of spontaneous deexcitation of this state, is of the order of 10^8 Hz (or 10^{-3} cm^{-1}).

1.2.4.2. Doppler Broadening. This type of broadening is caused by the thermal motion of the atoms of the gas. Due to the Doppler effect the resonance frequencies of different atoms differ. The width of the absorption line of a large number of the atoms of the gas, Γ_m, is determined by the temperature of the gas and in ordinary conditions ($T \approx 300$ K) is about 10^{11} Hz (or 1 cm^{-1}), which, as we see, is considerably greater than the natural broadening of the levels. If the external electromagnetic field impinges at right angles on a beam of atoms propagating in a certain direction, the linear Doppler effect vanishes, while the quadratic Doppler effect has a much smaller magnitude than the above estimate provides (v/c times less, where v is the atomic velocity; see Landau and Lifshitz, 1975, §48).

1.2.4.3. Ionization Broadening. One of the reasons for resonance broadening of a discrete state m is the possibility of this state being ionized by one photon from the external electromagnetic field. As noted earlier, in the optical range ω is lower than $\mathscr{E}_n$, the electron binding energy in the ground state. Hence, ionization (single-photon) broadening of the atomic ground states does not take place. Applying Eqs. (1.29) and (1.30) to the resonance state m, we find that

$$\Gamma_m = w_m = |z_{mk}|^2 E_\omega^2 p_k (4\pi^2)^{-1}. \tag{1.31}$$

Here k is the final state in the continuous spectrum in the transition $m \to k$. According to the law of energy conservation, $p_k^2/2 = \omega + \mathscr{E}_m$, with p_k the momentum of the electron in the finite state k in the continuous spectrum. Equation (1.31) shows that the role of single-photon ionization as a broadening mechanism increases with the intensity of the external electromagnetic field.

The single-photon ionization probability w_m also depends on the frequency ω of the electromagnetic field and the principal quantum number N_m (the binding energy) of the excited electronic state m. In the quasiclassical approximation ($N_m \gg 1$), $w_m \propto \omega^{-3} N_m^{-5}$ (see Sobelman, 1979, Section 9.5). We see that at a fixed frequency the probability drops sharply when N_m increases. But if $\hbar\omega \sim \mathscr{E}_m$, we find that $w_m \propto N_m$, since $\mathscr{E}_m = -1/2N_m^2$. Thus, in this case the single-photon ionization probability grows linearly with N_m.

One must bear in mind, however, that outside the framework of the model discussed here there are four additional mechanisms of resonance broadening.

1.2.4.4. Quadratic AC Stark Broadening in a Pulsed Field. In Section 1.1.4 we have seen that when the conditions (1.16) are met, the atomic levels experience a quadratic ac Stark shift. In the case at hand we are speaking of the shifts of resonance levels n and m, that is, $\delta\mathscr{E}_{n,\,m} \propto E_\omega^2$ [see Eq. (1.13)]. In a pulsed field, E_ω increases from zero to a maximum value, and then drops to zero. Correspondingly, the ac Stark shift of the resonance denominator $\omega_{mn} - \omega + \delta\mathscr{E}_m - \delta\mathscr{E}_n$ in the linear susceptibility changes from zero to a maximum value, and then back to zero. From the empirical viewpoint, a shift that varies in time in the described manner manifests itself as resonance broadening with a characteristic width $\Gamma_m \propto \delta\mathscr{E}_{m,n}$.

The situation becomes much more complex when the condition $z_{mn}E_\omega \ll \Delta = \omega_{mn} - \omega$, which is necessary for a quadratic Stark shift to occur in the resonance case, is not met. Below we analyze this case.

1.2.4.5. The Linear Stark Effect and Linear Stark Broadening in a Strong Electromagnetic Field. As the intensity of the electromagnetic field increases in the conditions of resonance, $\omega \approx \omega_{mn}$, the quadratic ac Stark effect becomes linear in E_ω. The linear shift in the resonance case is realized if

$$z_{mn}E_\omega \gg \Delta = \omega_{mn} - \omega. \qquad (1.32)$$

In this case the excited resonance state m is split into two states, known as *quasi energy states* (see Delone and Kraĭnov, 1985, Section 3.1), whose energies shift with respect to the unperturbed energies $\mathscr{E}_m$ by $\pm\Omega/2$, where the *resonance Rabi frequency* is (see Delone and Kraĭnov, 1985, Section 3.1)

$$\Omega = 2|z_{mn}E_\omega|. \qquad (1.33)$$

Note that the lower level n also splits into similar quasi energy states.

In a pulsed field, where the field is switched on and off, the magnitude of (1.33) varies in time from zero to a maximum value, and then back to zero. As in the case of quadratic Stark broadening, from the empirical viewpoint this appears as broadening of the resonance state m by $\Gamma_m \sim \Omega$. This is true when the time during which the field is switched on and off is small compared with the time it takes to register the effect.

1.2.4.6. Collision (Impact) Broadening. This broadening of atomic states is caused by the collisions of atoms with each other, as a result of which the atomic states become perturbed for a short time, and this perturbation leads to a shift and broadening of the levels. Naturally, the broadening depends strongly on the gas pressure. For instance, at a gas pressure of 1 Torr the typical values of Γ_m are 10^7–10^8 Hz (10^{-4}–10^{-3} cm^{-1}). We will ignore this type of broading in our discussion, since we are considering the model of a gas consisting of isolated atoms.

1.2.4.7. The Nonmonochromaticity of Laser Radiation. If the radiation interacting with the atoms of a gas is not strictly monochromatic but possesses a certain bandwidth $\Delta\omega$, this width (if large) can also influence the resonance width in the linear susceptibility. The same follows from the fact that the duration of a laser pulse is always finite: the inverse of the duration determines the resonance broadening.

1.2.5. Resonance Spontaneous Raman Scattering and Hot Luminescence

Let us consider the process in which under the action of an external electromagnetic field of frequency ω there occurs a resonance transition $n \to m$, that is, $\omega \approx \omega_{mn}$, after which (due to an infinitely weak field of frequency ν) an optical electron experiences a resonance transition $m \to p$, that is, $\nu \approx \omega_{mp}$. As noted in Section 1.2.4, state m splits into two quasi energy states. But which of the two states determines the resonance transition $n \to m \to p$? Note that this is a single-photon transition in the strong electromagnetic field of frequency ω and is therefore determined by the resonance linear susceptibility.

Let us denote the energies of the quasi energy levels of state m by $\mathscr{E}_m^+$ and $\mathscr{E}_m^-$ (Figure 1.7). As shown in Section 1.2.4, when the condition (1.32) is met, that is, in a very strong field, we have

$$\mathscr{E}_m^+ = \mathscr{E}_m + \frac{\Omega}{2}, \qquad \mathscr{E}_m^- = \mathscr{E}_m - \frac{\Omega}{2}. \qquad (1.34)$$

But if the field is weak, the energy of one of the two quasi energy levels becomes the energy of the unperturbed state m, that is, $\mathscr{E}_m^+ \to \mathscr{E}_m$. And the energy of the other quasi energy level, $\mathscr{E}_m^-$, becomes the energy of the first quasiharmonic component of the initial lower level n, that is, $\mathscr{E}_m^- \to \mathscr{E}_n + \hbar\omega$.

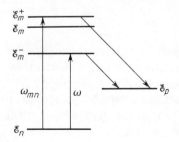

Figure 1.7. Two-photon transitions through the quasi energy states of an excited level under resonance excitation.

We start with the case of a weak field of frequency ω, when the condition opposite to (1.32) is met. The answer to the above question lies in the value of the ratio of the widths of the resonance levels m and n, or Γ_m/Γ_n (for more details see Rautian, 1975). Suppose that $\Gamma_m \gg \Gamma_n$, which takes place when, say, state n is the ground state. Then the width of the quasi energy level $\mathscr{E}_m^+$, equal to Γ_m (according to the Breit–Wigner procedure: $\mathscr{E}_m \to \mathscr{E}_m - i\Gamma_m$), proves to be much larger than that of the quasi energy level $\mathscr{E}_m^-$, which is equal to Γ_n (according to the Breit–Wigner procedure: $\mathscr{E}_n + \hbar\omega \to \mathscr{E}_n + \hbar\omega - i\Gamma_m$). Respectively, in conditions of resonance, the quantity $(\mathscr{E}_m^+ - \mathscr{E}_n - \hbar\omega)^{-1}$, which according to (1.22) determines the resonance linear susceptibility in terms of the quasi energy level $\mathscr{E}_m^+$, is of the order of Γ_m^{-1}, which is small compared with $(\mathscr{E}_m^- - \mathscr{E}_n - \hbar\omega)^{-1}$ and is of the order of Γ_n^{-1} (in conditions of resonance, of course). Thus, the transition proceeds generally through the quasi energy level $\mathscr{E}_m^-$, that is, essentially through the first quasiharmonic component of level n. Such a transition is known as a two-photon transition: a photon of frequency ω is absorbed, and a transition from the state with energy $\mathscr{E}_n$ into the state with energy $\mathscr{E}_n + \hbar\omega$ takes place. The transition from this state into the final state p, which is caused by a very weak field of frequency ν, can occur either when a photon ($\nu > 0$) of frequency ν is absorbed or when a photon ($\nu < 0$) of the same frequency is emitted. Thus, level m with an energy $\mathscr{E}_m$ becomes vacant in the process. Such a process is known as *resonance spontaneous Raman scattering*.

Similar reasoning shows that in the opposite limiting case, $\Gamma_m \ll \Gamma_n$, the resonance transition takes place through the quasi energy level $\mathscr{E}_m^+ \to \mathscr{E}_m$. Level m becomes populated in the process, and the transition is known as cascade (see Bloembergen, 1977). The entire process is called *hot luminescence*; it is similar to common luminescence, where a populated excited state of an atom is deexcited.

In a strong light field, where $|z_{mn}E_\omega| > \Delta$, the transition proceeds through both quasi energy levels, with interference terms present (Rautian, 1975), so that there is no way in which we can isolate a definite type of transition.

1.2.6. Conclusion

The following conclusion can be drawn from the above reasoning: the characteristics of all the processes containing resonance single-photon transitions can be expressed in terms of the resonance linear susceptibility. The various mechanisms of atomic-state broadening, which always exist, eliminate all nonphysical infinities which appear in the susceptibility when resonances are approached (see Section 1.1).

1.3. TRANSMISSION AND REFLECTION OF LIGHT AT NORMAL INCIDENCE TO AN INTERFACE

1.3.1. Introduction

The previous sections were devoted to the linear interaction of light with an isolated atom. Now we turn to a discussion of the main physical phenomena that take place when the light interacts linearly with gaseous media consisting of such atoms (the media are assumed to be rarefied). We devote this section to the question of how the strength of an electromagnetic field in a gaseous medium changes when the light impinges on the interface between a vacuum and the gaseous medium. The light wave is assumed to be strictly monochromatic with a plane wavefront. The interface is also assumed to constitute a plane.

An additional assumption used in this section is the absence of absorption when light passes through the medium. As we saw in Section 1.2, absorption plays an important role only when the frequency of the electromagnetic field and that of an atomic transition are in single-photon resonance; it is then determined by the imaginary part of the linear susceptibility. Absorption of light is considered in Section 1.4. Hence, here we assume that the atomic linear susceptibility $\chi^{(1)}$ is a real variable.

In the introduction to this chapter we noted that in the unperturbed-field approximation the variations of the strength of the electromagnetic field in a medium are small. This means that the linear polarization of the medium, defined as the average dipole moment per unit volume of the gas, can be calculated not from the value of the electromagnetic field in the medium (which itself requires calculating) but from the field of optical radiation impinging on the medium (the characteristics of which are assumed to be known). Below we give the criteria for applying the unperturbed-field approximation.

Equation (1.3) gives the polarization $\mathbf{P}^{(1)}(t)$ of one atom in an approximation that is linear in the field. This polarization is then averaged over the quantum-mechanical state of the atom. Hence, for the polarization vector of a unit volume of the gas we have a quantity that is N times greater, that is, $N\mathbf{P}^{(1)}(t)$. Here N is the number of atoms of the gas per unit volume. This statement, of course, is valid only for a rarefied gas, that is, when we can ignore the field induced by the dipole-dipole interaction between the atoms in comparison with the energy of an atomic dipole placed in the external electromagnetic field. It can be shown that this is equivalent to

$$N\chi^{(1)} \ll 1, \tag{1.35}$$

where $\chi^{(1)}$ is the linear atomic susceptibility.

Indeed, the dipole-dipole interaction of two neutral atoms is estimated as $d_1 d_2/r^3$, where r is the distance between the atoms, and d_1 and d_2 are the dipole moments of the atoms induced by the field. On the other hand, the energy of an atomic dipole in an external field is of the order of $d_1 E$. Obviously, the dipole-dipole interaction can be ignored if $d_1 d_2/r^3 \ll d_1 E$. Since $d_{1,2} \sim \chi^{(1)} E$ and $r^{-3} \sim N$, we arrive at (1.35). Note that if (1.35) is invalid, we are dealing with a condensed medium instead of a gas, since the distance between neighboring atoms will then be comparable to the atomic size. This last statement is strictly valid only when the susceptibility $\chi^{(1)}$ is nonresonance.

1.3.2. Maxwell's Equation in the Unperturbed-Field Approximation

Suppose $E'(t)$ is the strength of the electric field in the electromagnetic wave propagating in the medium. This vector satisfies the well-known Maxwell equation (see Born and Wolf, 1975)

$$\nabla^2 \mathbf{E}' - \frac{1}{c^2} \frac{\partial^2 \mathbf{E}'}{\partial t^2} = \frac{4\pi N}{c^2} \frac{\partial^2 \mathbf{P}^{(1)}}{\partial t^2}. \tag{1.36}$$

The right-hand side of this equation is written for a rarefied gas containing N atoms per unit volume.

The strength of the wave incident on the medium from a vacuum will be denoted by $\mathbf{E}(t)$. Assuming this field to be monochromatic, we can write $\mathbf{E}(t) \propto \exp(i\omega t)$. In the linear approximation the vectors $\mathbf{E}'(t)$ and $\mathbf{P}^{(1)}(t)$ depend on time in the same manner, that is,

$$\mathbf{E}'(t) \propto \exp(i\omega t), \qquad \mathbf{P}^{(1)}(t) \propto \exp(i\omega t). \tag{1.37}$$

With this in mind we can get rid of the time derivatives in Maxwell's equation (1.36):

$$\nabla^2 \mathbf{E}' + k^2 \mathbf{E}' = -4\pi N k^2 \mathbf{P}^{(1)}. \tag{1.38}$$

Here the wave number k is equal to ω/c.

The strength of the electric field in the electromagnetic wave in a vacuum,

$$\mathbf{E}(t) = \mathbf{E} \exp(i\omega t - i\mathbf{k} \cdot \mathbf{r}), \tag{1.39}$$

obviously obeys (1.38) without the right-hand side:

$$\nabla^2 \mathbf{E} + k^2 \mathbf{E} = 0. \tag{1.40}$$

This can easily be verified directly if we substitute the expression (1.39) for the field strength in a monochromatic plane wave into Eq. (1.40); the latter becomes an identity as a result.

Let us introduce the vector $\delta\mathbf{E} = \mathbf{E}' - \mathbf{E}$, which is the variation of the electromagnetic field strength in the medium in relation to that in the vacuum. This vector can also be called the electric field strength in the electromagnetic wave due to linear polarization in the medium. The equation for $\delta\mathbf{E}$ can be found by subtracting (1.40) from (1.38):

$$\nabla^2 \delta\mathbf{E} + k^2 \delta\mathbf{E} = -4\pi N k^2 \mathbf{P}^{(1)}. \tag{1.41}$$

The same equation, obviously, holds for the Fourier transforms with respect to time of the quantities $\delta\mathbf{E}$ and $\mathbf{P}^{(1)}$ at frequency ω. Hence, we can rewrite Eq. (1.41) in the following form (for the time being, the exact form):

$$\nabla^2 \delta\mathbf{E} + k^2 \delta\mathbf{E} = -4\pi N k^2 \chi^{(1)}(\omega; \omega)\mathbf{E}'. \tag{1.42}$$

Here $\chi^{(1)}(\omega; \omega)$ is the linear susceptibility of an atom at frequency ω (see Section 1.1). In what follows we will usually drop the independent variables in $\chi^{(1)}$ for the sake of brevity.

The essence of the unperturbed-field approximation is that into the right-hand side of Maxwell's equation (1.42) we substitute not the true field $\mathbf{E}'$ in the medium but the field $\mathbf{E}$ in the vacuum. Obviously, the condition of applicability of such an approximation is simple:

$$\delta\mathbf{E} \ll \mathbf{E}. \tag{1.43}$$

Below we will see that if (1.43) is met, the basic variation of the field strength in the medium lies in the phase of the field rather than its absolute value, while for the reflected wave the opposite is true. After such a substitution, Eq. (1.42) assumes the form of inhomogeneous linear differential equation with a known right-hand side:

$$\nabla^2 \delta\mathbf{E} + k^2 \delta\mathbf{E} = -4\pi N k^2 \chi^{(1)}\mathbf{E}$$
$$= -4\pi N k^2 \chi^{(1)}\mathbf{E}_\omega \exp(i\omega t - i\mathbf{k} \cdot \mathbf{r}). \tag{1.44}$$

1.3.3. The Wave of Linear Polarization of a Medium

Let us orient the z axis along the wave vector $\mathbf{k}$ of the electromagnetic wave incident on the medium from the vacuum. The medium-vacuum interface is the plane $z = 0$, in accordance with the assumption that the light propagates at normal incidence to the interface (Figure 1.8). The vector $\mathbf{E}$ is

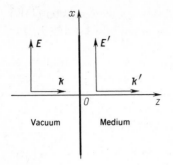

Figure 1.8. Field strengths and wave vectors at normal incidence of a light wave on the interface $z = 0$ between vacuum and medium.

directed along the x axis. Then Eq. (1.44) implies that the strength of the polarization wave, $\delta \mathbf{E}$, is also directed along the x axis. Let us find the magnitude of $\delta \mathbf{E}$.

The solution to Eq. (1.44) that satisfies the boundary condition $\delta \mathbf{E}(z = 0) = 0$ (the absence of a polarization wave at the interface) and ignores the reflected wave (see Section 1.3.5) in comparison with $\delta \mathbf{E}$ (and the more so in comparison with $\mathbf{E}$) can be found immediately:

$$\delta \mathbf{E} = -2\pi i N \chi^{(1)} k z \mathbf{E}_\omega \exp(i\omega t - ikz). \qquad (1.45)$$

We see that the strength of the polarization wave grows linearly with z when light is passing through the medium.

The condition of applicability of the solution (1.45), that is, the validity of the unperturbed-field approximation (1.43), is obtained by simply substituting (1.45) into (1.43). The result is

$$N\chi^{(1)}\frac{z}{\lambda} \ll 1. \qquad (1.46)$$

Here $\lambda = k^{-1}$ is the wavelength of the electromagnetic field incident on the medium. We see that for (1.46) to be valid in a real situation, when $z \gg \lambda$ (that is, with the dimensions of the medium large compared to the wavelength) we must ensure that (1.35) is satisfied with a large margin. Note that at $z \sim \lambda$, approximating electromagnetic waves by light rays loses all meaning.

The total field $\mathbf{E}'$ in the medium, with allowance made for (1.45) and (1.39), can be written as follows:

$$\mathbf{E}' = \left[1 - 2\pi i N \chi^{(1)} \mathbf{k} \cdot \mathbf{r}\right]\mathbf{E}_\omega \exp(i\omega t - i\mathbf{k} \cdot \mathbf{r}). \qquad (1.47)$$

1.3.4. The Dielectric Constant of a Linear Medium

The expression within the square brackets in (1.47) can be rewritten within the same approximation, that is, when (1.46) is met, as

$$\mathbf{E}' = \mathbf{E}_\omega \exp(i\omega t - i\mathbf{k}' \cdot \mathbf{r}), \tag{1.48}$$

where we have introduced the notation

$$\mathbf{k}' = \left(1 + 2\pi N\chi^{(1)}\right)\mathbf{k}. \tag{1.49}$$

Thus, we see that when an electromagnetic wave enters the medium, the maximum $\mathbf{E}_\omega$ of the electric field strength does not change: the only change is associated with the wave vector, which changes a little in length but not in direction. Of course, this statement is valid only if we ignore the small reflected wave (see Section 1.3.5).

If we introduce the index of refraction n_ω of the medium at the field frequency ω through the usual relationship,

$$k' = n_\omega \frac{\omega}{c}, \tag{1.50}$$

then, after substituting (1.50) into (1.49), we obtain

$$n_\omega = 1 + 2\pi N\chi^{(1)}(\omega; \omega). \tag{1.51}$$

From the phenomenological Maxwell equations in a medium we can obtain a relation that connects n_ω with the dielectric constant ϵ_ω of the medium, the latter being the proportionality factor between the electric induction $\mathbf{D}_\omega$ and the electric field strength $\mathbf{E}_\omega$:

$$n_\omega = \epsilon_\omega^{1/2}. \tag{1.52}$$

In writing this relationship we assume the permeability of the gaseous medium to be unity. Combining (1.52) and (1.51) and taking into account the inequality (1.35), we find that

$$\epsilon_\omega = 1 + 4\pi N\chi^{(1)}(\omega; \omega). \tag{1.53}$$

We have therefore established that both the index of refraction n_ω and the dielectric constant ϵ_ω of a gaseous medium differ little from unity.

Note that Eq. (1.53) is valid even if the condition (1.35) is not met, that is, if ϵ_ω is considerably greater than unity (see Born and Wolf, 1975).

Consequently, if instead of (1.49) we write

$$\mathbf{k}' = \left(1 + 4\pi N\chi^{(1)}\right)^{1/2}\mathbf{k}, \tag{1.54}$$

then the solution (1.48) for $\mathbf{E}'$ with such a vector $\mathbf{k}'$ will be an exact solution of Maxwell's equation (1.38), irrespective of the unperturbed-field approximation. This can be easily verified. Substituting (1.48) into (1.38), we obtain

$$\left(k^2 - k'^2\right)\mathbf{E}' = -4\pi Nk^2\mathbf{P}^{(1)}$$
$$= -4\pi Nk^2\chi^{(1)}(\omega; \omega)\mathbf{E}'. \tag{1.55}$$

Canceling $\mathbf{E}'$, we arrive at the following relationship:

$$k'^2 = k^2\left(1 + 4\pi N\chi^{(1)}\right), \tag{1.56}$$

which obviously coincides with (1.54).

1.3.5. Linear Reflection of Light from a Medium

As the solution (1.48) shows, the intensity $I' = c|\mathbf{E}'|^2/2\pi$ of a wave that propagates through a medium is equal to the intensity $I = c|\mathbf{E}|^2/2\pi$ of the wave in vacuum, since the only change in (1.48) is that in the phase of the wave. Thus,

$$I' = I = \frac{c|\mathbf{E}_\omega|^2}{2\pi}. \tag{1.57}$$

Actually, this is not quite so, since there is always a small (but finite) reflection of the electromagnetic wave from the interface between the vacuum and the gaseous medium. Our aim now is to determine the intensity of the reflected wave.

As we will see later, $|\mathbf{E}'|$, which determines the intensity I' of the wave in the medium, differs from $|\mathbf{E}|$, which determines the intensity I of the wave in vacuum, by a factor of the order of $N\chi^{(1)}E \ll E$. But if we turn to (1.47), we will see that $\mathbf{E}'$ differs from $\mathbf{E}$ by a factor of the order of $\delta\mathbf{E} \sim N\chi^{(1)}kz\mathbf{E}$ [see Eq. (1.45)]. This is much larger than $N\chi^{(1)}E$, since we assume that $kz \gg 1$ (that is, $z \gg \lambda$; see the end of Section 1.3.3). No wonder then that we lost the much smaller reflected wave in determining the small polarization wave $\delta\mathbf{E} \ll \mathbf{E}$. But how are we to restore and determine the field strength and intensity of the reflected wave?

Let us denote by $\mathbf{E}''$ the electric field strength of the wave reflected from the medium and propagating in the direction $z < 0$ opposite to that of the

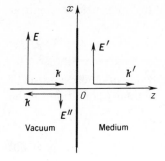

Figure 1.9. Field strengths and wave vectors at normal incidence of a light wave on the interface $z = 0$ between vacuum and medium; the reflected wave is taken into account.

incident wave (Figure 1.9). Then we can write

$$\mathbf{E}'' = \mathbf{E}''_\omega \exp(i\omega t + ikz). \tag{1.58}$$

If we compare this with (1.39), we will see that the wave vector of the reflected wave differs in direction from the wave vector **k** of the incident wave but coincides in magnitude. This follows from the fact that $\mathbf{E}''$, like $\mathbf{E}$, satisfies the Maxwell equation without the right-hand side, (1.40), that is, in vacuum, where polarization does not occur.

As for the direction of $\mathbf{E}''$, it is clear that this vector is directed along the x axis (the same is true for $\mathbf{E}$ and for $\mathbf{E}'$, which is the electric field in the electromagnetic wave propagating in the medium). This implies that all relationships involving these quantities can be written in scalar form.

At the vacuum-medium interface ($z = 0$) the right-hand side of Maxwell's equation (1.38) experiences a jump, since $\mathbf{P}^{(1)}$ experiences a jump in the z direction from zero to the finite value that it assumes in the medium. If we turn to Eq. (1.38), we can conclude that the second derivative of the field strength with respect to z (that is, $\partial^2\mathbf{E}'/\partial z^2$, which enters into the Laplacian operator) experiences a jump. If we integrate both sides of Eq. (1.38) with respect to z over an infinitely small region in the vicinity of the interface at $z = 0$, then on the right-hand side of (1.38) we have zero. Hence, the first derivative $\partial\mathbf{E}'/\partial z$, which appears after integration of $\partial^2\mathbf{E}'/\partial z^2$ with respect to z on the left-hand side, must be continuous at the interface.

Moreover, integrating both sides of (1.38) with respect to z once more, we find that the field strength $\mathbf{E}'$ too must be continuous at the interface. Note that the presence of the second term, $k^2\mathbf{E}'$, on the left-hand side of (1.38) changes nothing in the above reasoning, because of the continuity of the field strength.

Thus, the boundary continuity condition for the total field strength at $z = 0$ has the form

$$\mathbf{E}_\omega + \mathbf{E}_\omega'' = \mathbf{E}_\omega' \qquad (1.59)$$

[the total field, which corresponds to the left-hand side of (1.59), consists of the incident and reflected fields]. Moreover, the continuity condition for the first derivatives,

$$\frac{\partial \mathbf{E}}{\partial z} + \frac{\partial \mathbf{E}''}{\partial z} = \frac{\partial \mathbf{E}'}{\partial z}, \qquad (1.60)$$

with

$$\begin{aligned}
\mathbf{E} &= \mathbf{E}_\omega \exp(i\omega t - ikz), \\
\mathbf{E}' &= \mathbf{E}_\omega' \exp(i\omega t - ik'z), \\
\mathbf{E}'' &= \mathbf{E}_\omega'' \exp(i\omega t + ikz),
\end{aligned} \qquad (1.61)$$

assumes the form where k' is defined in (1.56)

$$k(\mathbf{E}_\omega - \mathbf{E}_\omega'') = k'\mathbf{E}_\omega'. \qquad (1.62)$$

To find the field strength in the reflected wave, $\mathbf{E}_\omega''$, we must exclude $\mathbf{E}_\omega'$ from (1.59) and (1.62). This is done by dividing (1.62) by (1.59) termwise:

$$k(\mathbf{E}_\omega - \mathbf{E}_\omega'') = k'(\mathbf{E}_\omega + \mathbf{E}_\omega''). \qquad (1.63)$$

Bearing in mind that $E_\omega'' \ll E_\omega$ and substituting the expression (1.56) or (1.49) for k' into (1.63), we arrive at

$$1 - \frac{2E_\omega''}{E_\omega} = 1 + 2\pi N\chi^{(1)}, \qquad (1.64)$$

whence

$$E_\omega'' = -\pi N\chi^{(1)}(\omega; \omega)E_\omega. \qquad (1.65)$$

This relation was used at the beginning of this section to estimate E''.

The *reflection coefficient* R_ω is defined as the ratio of the intensity of the reflected wave to that of the incident wave:

$$\begin{aligned}
R_\omega &= \left| \frac{E_\omega''}{E_\omega} \right|^2 \\
&= \left| \pi N\chi^{(1)}(\omega; \omega) \right|^2 \ll 1. \qquad (1.66)
\end{aligned}$$

We see that only a small fraction of the electromagnetic wave intensity is reflected from the interface separating the vacuum from the rarefied gas. The magnitude of R_ω becomes noticeable only for condensed media, when $N\chi^{(1)}$ is of the order of unity.

1.3.6. Conclusion

Both the change experienced by an electromagnetic wave in a medium and the reflection of the wave from the interface are determined in linear optics by the linear atomic susceptibility $\chi^{(1)}(\omega; \omega)$. The index of refraction and the dielectric constant of the medium at frequency ω are expressed in terms of this susceptibility. The only quantity that characterizes a rarefied gaseous medium in the nonresonance case is the number of atoms of the gas per unit volume, N.

1.4. LINEAR ABSORPTION OF LIGHT

1.4.1. Introduction

In Section 1.3 it was assumed that the linear susceptibility is of a nonresonance nature. As the results of Section 1.2 show, if single-photon resonances $\omega \approx \omega_{mn}$ between the frequency of the external electromagnetic field with the frequency of an allowed bound-bound atomic transition are realized in the linear susceptibility, then $\chi^{(1)}(\omega; \omega)$ becomes complex-valued.

We immediately note that this is impossible for states m in the continuous spectrum, since in Section 1.2.3 it was found that single-photon ionization from the ground state of an atom, which according to (1.29) results in $\chi^{(1)}$ having an imaginary part, does not occur in the optical frequency range.

However, ionization is possible from an excited state m of the atom (see Section 1.2.4.3), which results in the resonance level m having a nonzero ionization width (1.31) ($\omega_{mn} \approx \omega$). The photons from the external electromagnetic field leave the light beam, which leads to a drop in the beam's intensity as the light passes through the gaseous medium. This constitutes one of the mechanisms of light absorption.

Another mechanism, which manifests itself during resonance excitation of a discrete state m, is caused by spontaneous deexcitation of this state through emission of a cascade of photons of various energies (see Section 1.2.4.1) while photons from the incident light beam disappear.

If we are dealing with Rayleigh scattering, the direction in which a spontaneous photon is emitted may be arbitrary (although this photon is of the same frequency as the photon from the external electromagnetic field absorbed by an atom), while the direction of an external electromagnetic field photon is fixed (it flies along the z axis). Thus, the emerging scattering of light of the same frequency leads to a drop in the intensity of the light beam propagating along the z axis, that is linear light absorption. It is obvious that the total number of photons propagating in all directions remains the same in the process.

Of course, a certain amount of light is absorbed by the medium even if the resonance is out of tune, down to a completely nonresonance case. But in the nonresonance case the imaginary part of the linear susceptibility is extremely small. For this reason we will consider only the resonance case.

1.4.2. The Absorption Coefficient

The formula (1.48) for the electric field strength in the electromagnetic wave propagating in a medium remains valid even if the linear susceptibility $\chi^{(1)}(\omega; \omega)$ becomes complex-valued (the resonance case). Substituting (1.49) into (1.48), we can write the field strength as follows:

$$\mathbf{E}' = \mathbf{E}_\omega \exp\left[i\omega t - ikz\left(1 + 2\pi N\,\mathrm{Re}\,\chi^{(1)} + 2\pi Ni\,\mathrm{Im}\chi^{(1)}\right)\right]. \quad (1.67)$$

Here we have ignored the small changes in the intensity of the transmitted wave caused by reflection from the interface (see Section 1.3.5). In (1.67) $\chi^{(1)}$ is determined by a relationship of the type (1.25) when an atom emits a photon of frequency ν after resonantly absorbing a photon from the external electromagnetic field of frequency ω.

The imaginary part of $\chi^{(1)}$ in (1.67) leads to an exponential decrease in the electric field strength as z grows, with an exponent that does not depend on the light intensity (Bouguer's law); if

$$I'(z) = I'(0)\exp(-\kappa z), \quad (1.68)$$

then κ is the absorption coefficient for an electromagnetic wave propagating in a medium, and κ^{-1} is the characteristic length over which the intensity of the incident light drops e-fold. Combining (1.67) and (1.68) with the fact that $I' \propto |\mathbf{E}'|^2$, we find that

$$\kappa = -4\pi Nk\,\mathrm{Im}\,\chi^{(1)}(\omega; \omega), \quad (1.69)$$

which implies that κ is independent of E'. The fact that κ is proportional to N is known as *Beer's law*. The above formula forces us to conclude that the imaginary part of the susceptibility is negative, since $\kappa > 0$.

If we substitute (1.22) into (1.69) and consider, for the sake of simplicity, the case of an exact resonance, we find the following estimate:

$$\kappa \approx 4\pi N k \Gamma_m^{-1}(\mathbf{r}_{mn} \cdot \mathbf{e})^2. \qquad (1.70)$$

Far from a resonance the value of κ rapidly drops. Substituting (1.22) into (1.69), we arrive at the following estimate in this case:

$$\kappa \approx 4\pi N k \Gamma_m (\mathbf{r}_{mn} \cdot \mathbf{e})^2 (\omega_{mn} - \omega)^{-2}. \qquad (1.71)$$

In Eqs. (1.70) and (1.71), just as before, Γ_m stands for the width of the resonance. Depending on the situation, the quantity Γ_m is determined within the framework of the model considered either by the probability of single-photon ionization out of an excited state m or by spontaneous relaxation of this state, with the latter resulting in either Rayleigh scattering or spontaneous Raman scattering. Consequently, to calculate Γ_m we must use Eq. (1.31) for ionization, (1.17) for Rayleigh scattering, and (1.20) for spontaneous Raman scattering. In a real situation we must consider the mechanisms of resonance broadening considered in Section 1.2.4.

1.4.3. Conclusion

The main idea of this section is that within the framework of linear optics the absorption of light is determined by the linear susceptibility $\chi^{(1)}$, which is a characteristic of the substance (medium) and does not depend on the intensity of the light. This fact results in Bouguer's and Beer's laws.

1.5. REFRACTION OF AN ELECTROMAGNETIC WAVE FALLING AT AN ANGLE ON THE INTERFACE BETWEEN TWO MEDIA

1.5.1. Introduction

In Section 1.3 we considered normal incidence of an electromagnetic wave at a boundary separating a vacuum from a gaseous medium. Here we will investigate the case where the wave vector $\mathbf{k}$ of the incident wave forms an angle α with the normal to the interface $z = 0$. The aim of this section is to show how a weak linear polarization wave emerges in the medium in the unperturbed-field approximation. We will find that this wave is added to the incident wave and the two form a total wave whose wave vector differs somewhat in direction from the wave vector **k** of the incident wave. The phenomenon is known as refraction. It is obvious, of course, that for a gaseous medium this change in direction is small compared to α.

As noted in Section 1.3.5, in addition to the weak linear-polarization wave there appears a reflected electromagnetic wave. But, as we found out, at $z \gg \lambda$ the effect of this reflected wave on the polarization wave can be ignored. For this reason the following exposition will be similar to that of Sections 1.3.2 and 1.3.3, that is, we ignore the presence of a reflected wave.

1.5.2. Maxwell's Equation in the Unperturbed-Field Approximation

Let us denote, as we did before, the low electric field in the polarization wave as δE. Maxwell's equation for this quantity in the unperturbed-field approximation has the form (1.44). We also assume, as before, that the z axis points in the direction normal to the interface between vacuum and medium. For the sake of simplicity we assume the incident wave to be linearly polarized (as we did above). Finally, we direct the electric field strength **E** in this wave along the x axis.

In contrast to the situation in the previous section, the wave vector **k** of the incident electromagnetic wave lies in the yz plane and forms an angle with the normal to the interface (Figure 1.10; this angle, as noted above, we denote by α). The wave vector of the polarization wave lies in the same plane but is directed differently. The transverse nature of electromagnetic fields implies that δE must be perpendicular to the direction of the wave vector of the polarization wave. Hence, both **E** and δE are directed along the x axis.

Let us rewrite Eq. (1.44) in scalar form, isolating the variables y and z on which the electric field strength in the polarization wave depends:

$$\frac{\partial^2 \delta E(y, z)}{\partial y^2} + \frac{\partial^2 \delta E(y, z)}{\partial z^2} + k^2 \delta E(y, z)$$
$$= -4\pi N k^2 \chi^{(1)} E_\omega \exp\left(i\omega t - ik_y y - ik_z z\right). \quad (1.72)$$

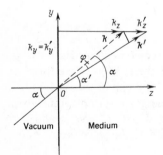

Figure 1.10. Wave vectors of the incident and refracted electromagnetic waves falling at an angle to the interface between vacuum and medium.

Here k_y and k_z are the projections of vector $\mathbf{k}$ on the respective coordinate axes, so that $k^2 = k_y^2 + k_z^2$.

The solution to Eq. (1.72) that satisfies the boundary condition specifying the absence of a polarization wave at the interface between vacuum and medium, that is, at $z = 0$, can be found directly [see (1.45) in this connection]:

$$\delta E(y, z) = -2\pi i N \chi^{(1)} k^2 k_z^{-1} z E \omega \exp(i\omega t - ik_y y - ik_z z). \quad (1.73)$$

Adding the polarization wave (1.73) to the incident electromagnetic wave, we find the overall electric field strength in the medium, $\mathbf{E}' = \mathbf{E} + \delta\mathbf{E}$:

$$\mathbf{E}'(y, z) = \left[1 - 2\pi i N \chi^{(1)} k^2 k_z^{-1} z\right] \mathbf{E}_\omega \exp(i\omega t - ik_y y - ik_z z). \quad (1.74)$$

When the light falls on the medium at right angles, this expression leads to (1.47), as expected.

Just as we went over from (1.47) to (1.48) by rejecting the unperturbed field approximation, we can go over from (1.74) to

$$\mathbf{E}'(y, z) = \exp\left[-2\pi i N \chi^{(1)} k^2 k_z^{-1} z\right] \mathbf{E}_\omega \exp(i\omega t - ik_y y - ik_z z). \quad (1.75)$$

We can rewrite this as follows:

$$\mathbf{E}'(y, z) = \mathbf{E}_\omega \exp(i\omega t - ik_y y - ik_z' z), \quad (1.76)$$

where we have introduced the notation

$$k_z' = k_z + 2\pi N \chi^{(1)} k^2 k_z^{-1} \quad (1.77)$$

[see (1.49) in this connection].

1.5.3. The Turning of the Wave Vector in a Medium

Let us analyze the solution (1.76), (1.77) we obtained for the electromagnetic wave in the medium. First, we see that the wave number k_y does not change as we go from vacuum to medium. From the physical standpoint this is explained by the homogeneity of the system considered in the y direction. On the other hand, k_z changes to k_z'. In terms of geometrical optics this means that the direction of propagation of the electromagnetic wave (specified by the angle α) turns when the wave enters the medium. Let us estimate the turn angle φ. The geometry of weak refraction of a light beam by a medium is shown in Figure 1.10. There α' is the angle between the normal to the interface and the wave vector $\mathbf{k}'$ of the light ray in the medium. Obviously, $\varphi = \alpha - \alpha'$.

By the definition of α we have $\sin \alpha = k_y/k$. On the other hand, from Figure 1.10 we see that φ and α are related thus:

$$k\varphi = (k_z' - k_z)\sin \alpha. \tag{1.78}$$

From this we get

$$k^2\varphi = (k_z' - k_z)k_y. \tag{1.79}$$

Combining this with (1.77), we find that

$$\varphi = 2\pi N\chi^{(1)}\frac{k_y}{k_z}. \tag{1.80}$$

Naturally, at normal incidence ($\alpha = 0$) Eq. (1.80) yields $\varphi = 0$. At an arbitrary angle of incidence α, when $k_y \sim k_z \sim k$, the fact that $\varphi \ll 1$ (i.e. $\varphi \ll \alpha$) follows, as shown by (1.80), from $N\chi^{(1)}$ being much less than unity [see (1.35)], and $k_z' - k_z \ll k_z$, according to (1.77).

Equation (1.80) can easily be related to the well-known Snell's law (see the Introduction to Chapter 1),

$$\frac{\sin \alpha}{\sin \alpha'} = n_\omega = \epsilon_\omega^{1/2}. \tag{1.81}$$

Here ϵ_ω is the dielectric constant of the gaseous medium, and n_ω is the index of refraction of this medium. In the case at hand, $\alpha' = \alpha - \varphi$, with $\varphi \ll \alpha$, so that the left-hand side of (1.81) can be rewritten thus: $1 + \varphi \cot \alpha$. As for the right-hand side, according to (1.51) we can write it as follows: $1 + 2\pi N\chi^{(1)}(\omega; \omega)$. Hence, from (1.81) we find that

$$\varphi = 2\pi N \tan \alpha \, \chi^{(1)}(\omega; \omega). \tag{1.82}$$

Since $k_y = k_z \tan \alpha$, Eqs. (1.82) and (1.80) coincide, as expected.

We can examine the refraction process from a somewhat different angle. We direct the wave vector $\mathbf{k}$ of the incident electromagnetic wave along the z axis. Then the interface separating the vacuum from the medium will depend on the position of point y along the wavefront. The result is a dependence of the electromagnetic properties of the medium on the y coordinate. Thus, in accordance with the above, we can conclude that if the medium is inhomogeneous with respect to the transverse coordinate y along the z axis, then the wave changes its line of propagation, that is, there appears a small component of the electromagnetic field along the y axis.

As for the reflected wave, Maxwell's equation for this wave immediately yields $k_y'' = k_y$ and $k_z'' = -k_z$. We have thus arrived at the well-known law of reflection: the angle of incidence equals the angle of reflection. The above relationships follow from the absence of polarization in a vacuum, that is, at $z < 0$.

1.5.4. The Falling of a Spatially Inhomogeneous Light Beam on a Linear Medium

The inhomogeneity with respect to y, of which we spoke in Section 1.5.3, may appear not only when the light falls at an angle to the interface but also when the electric field strength **E** of the incident electromagnetic wave depends on the transverse coordinate y. For instance, if we are dealing with a spatially inhomogeneous light beam whose intensity vanishes far from the beam axis, then in this realistic situation the electric field strength will depend on the transverse coordinate r of the cylindrical system r, z, ψ. But for the sake of mathematical simplicity we remain within the framework of the Cartesian system of coordinates and assume that **E** depends only on y, while remaining independent of x.

How does refraction occur for such a wave with a simple dependence of **E** on y? Suppose that the electric field strength in a vacuum satisfies Maxwell's equation

$$\frac{\partial^2 \mathbf{E}}{\partial y^2} + \frac{\partial^2 \mathbf{E}}{\partial z^2} + k^2 \mathbf{E} = 0 \qquad (1.83)$$

and the second Maxwell equation,

$$\frac{\partial E_y}{\partial y} + \frac{\partial E_z}{\partial z} = 0, \qquad (1.84)$$

with $k = \omega/c$. Equation (1.84) is satisfied if we orient the x axis along the vector **E**. Then $E_y = E_z = 0$. And Eq. (1.83) is satisfied if we express its solution, as can easily be verified, in the form

$$E = E_x = E_\omega \left(1 - \frac{y}{a}\right) \exp(-ikz). \qquad (1.85)$$

Here a is the distance over which the electric field strength varies in the transverse direction y, or roughly the size of the light beam. Note that in all these formulas we have taken for granted the factor $\exp(i\omega t)$, giving the time dependence of the electric field strength.

Hence, suppose that an electromagnetic wave of the type (1.85) is falling from a vacuum on a gaseous medium normally, with the interface between vacuum and medium being the plane $z = 0$. Our aim is to determine the electric field strength in the medium. Maxwell's equation in the unperturbed-field approximation has the form [see (1.44)]

$$\frac{\partial^2 \delta E(y, z)}{\partial y^2} + \frac{\partial^2 \delta E(y, z)}{\partial z^2} + k^2 \delta E(y, z)$$

$$= -4\pi Nk^2 \chi^{(1)} E_\omega \left(1 - \frac{y}{a}\right) \exp(-ikz). \quad (1.86)$$

Here $\delta E(y, z)$ is the electric field strength in the electromagnetic wave of linear polarization in the medium, and in writing the right-hand side we have used the formula (1.85) for the electric field strength in the incident electromagnetic wave.

The solution to Eq. (1.86) can be found immediately:

$$\delta E(y, z) = -2\pi ikz N\chi^{(1)} E_\omega \left(1 - \frac{y}{a}\right) \exp(-ikz), \quad (1.87)$$

where we have used the boundary condition $\delta E(y, z = 0) = 0$. As we can easily see, in the absence of a dependence on the transverse size a, (1.87) transforms into (1.45).

We see that the vector δE, just like E, is directed along the x axis. Adding (1.87) and (1.85), we arrive at an expression for the total electric field in the electromagnetic wave in the medium:

$$E'(y, z) = \left(1 - 2\pi iN\chi^{(1)}kz\right) E_\omega \left(1 - \frac{y}{a}\right) \exp(-ikz). \quad (1.88)$$

In a manner similar to that used in going over from (1.47) to (1.48), we can rewrite the solution (1.88) without resorting to the unperturbed-field approximation:

$$E'(y, z) = E_\omega \left(1 - \frac{y}{a}\right) \exp(-ik'z), \quad (1.89)$$

with

$$k' = \left(1 + 2\pi N\chi^{(1)}\right)k, \quad (1.90)$$

which coincides with (1.49). Thus, the solution (1.89) enables us to conclude that the linear polarization of a medium does not change the spatial inhomogeneity in the distribution of intensity in the light beam. The change in the wave number in the medium from that in the vacuum defined by (1.90) proves to be the same as for the plane, infinite wavefront of a monochromatic electromagnetic wave. As the above conclusion implies, a wavefront with any degree of inhomogeneity remains undistorted in a linear medium, due to the arbitrariness of a. In Chapter 3 we will see that the situation changes drastically in a nonlinear medium, where light focusing and other effects emerge.

1.5.5. Conclusion

Usually Snell's law for a linear medium is derived within the framework of geometrical linear optics for two rays, the incident and the refracted. The aim of the present section was to derive this law within the framework of wave optics, since nonlinear light refraction in a gaseous medium (see Chapter 3) will be considered along these lines. Moreover, we have proved that in a linear medium there are no distortions of a spatially inhomogeneous light beam.

1.6. POLARIZATION OF AN ELECTROMAGNETIC FIELD IN A LINEAR MEDIUM

1.6.1. Introduction

The word "polarization" is used in two meanings in this book. If we speak of atomic polarization, we are dealing with the average dipole moment of an atom induced by the electromagnetic field. On the other hand, the polarization of an electromagnetic field is the direction of the electric field vector in the electromagnetic wave. In this section we will see how the polarization of an electromagnetic field changes when the electromagnetic wave travels in a linear gaseous medium. Only completely polarized light will be considered. An arbitrary partially polarized electromagnetic wave can be represented as a linear combination of two elliptically polarized waves. The parameters that characterize partially polarized light are known as *Stokes parameters*. For completely polarized light the Stokes parameters assume definite values (see Landau and Lifshitz, 1975). They are related to the angles characterizing completely polarized light, introduced below, through well-known formulas, which we do not give here (see Landau and Lifshitz, 1975, §50).

1.6.2. Parameters Characterizing an Elliptically Polarized Electromagnetic Wave

Take the most general case of a completely polarized electromagnetic wave that falls on a linear medium. The real-valued electric field **E** in such a wave can be written in the form

$$\mathbf{E} = \tilde{\mathbf{E}} + \tilde{\mathbf{E}}^*. \tag{1.91}$$

The vector $\tilde{\mathbf{E}}$ corresponds to a monochromatic electromagnetic wave of frequency ω and wave vector **k** (the length of this vector is $k = \omega/c$) that travels along the z axis:

$$\tilde{\mathbf{E}} = \mathbf{E}_\omega \exp(i\omega t - ikz). \tag{1.92}$$

Generally speaking, the complex-valued vector $\mathbf{E}_\omega$ lies in the xy plane perpendicular to the direction of propagation of the electromagnetic wave, z. This vector can be resolved into two basis vectors in the xy plane, for which it is convenient to take not the unit vectors $\mathbf{e}_x$ and $\mathbf{e}_y$ along the x and y axes, respectively, but the linear combinations

$$\mathbf{e}_+ = -2^{-1/2}(\mathbf{e}_x + i\mathbf{e}_y), \qquad \mathbf{e}_- = 2^{-1/2}(\mathbf{e}_x - i\mathbf{e}_y). \tag{1.93}$$

(These new basis vectors constitute the so-called spherical system of coordinates.) Thus, in the general case we have

$$\mathbf{E}_\omega = A\mathbf{e}_+ + B\mathbf{e}_-, \tag{1.94}$$

where A and B are complex-valued expansion coefficients.

An elliptically polarized wave is fixed by the coefficients in (1.94), specifically

$$\mathbf{E}_\omega = E_\omega[\cos(\theta + \pi/4)\exp(i\varphi)\,\mathbf{e}_+$$
$$+ \cos(\theta - \pi/4)\exp(-i\varphi)\,\mathbf{e}_-]. \tag{1.95}$$

What is the physical meaning of the angles θ and φ in (1.95)? Let us first show that $\tan\theta$ defines the ratio of the semimajor axes in the polarization ellipse. Let us find the square of the absolute value of the electric field strength:

$$|\mathbf{E}|^2 = |\tilde{\mathbf{E}} + \tilde{\mathbf{E}}^*|^2$$
$$= \frac{E_\omega^2}{2}[1 - \cos 2\theta \cos(2kz - 2\omega t)]. \tag{1.96}$$

In deriving this relationship we used the easily varifiable orthogonality conditions for unit vectors $\mathbf{e}_+$ and $\mathbf{e}_-$:

$$\mathbf{e}_+ \cdot \mathbf{e}_- = 1, \qquad \mathbf{e}_+ \cdot \mathbf{e}_+ = \mathbf{e}_- \cdot \mathbf{e}_- = 0. \tag{1.97}$$

The maximum value of $|E|^2$ is attained, according to (1.96), at $2kz - 2\omega t = \pi$ and is

$$|E|^2_{max} = E^2_\omega \cos^2\theta. \tag{1.98}$$

The minimum value $|E|^2_{min}$ can be found in a similar way. It is attained at $2kz - 2\omega t = 0$ and is

$$|E|^2_{min} = E^2_\omega \sin^2\theta. \tag{1.99}$$

Note that the maximum and minimum values change places for $|\theta| > \pi/4$.

Equations (1.98) and (1.99) represent the squares of the semimajor axes of the polarization ellipse in an elliptically polarized electromagnetic wave. We see that the ratio of the semiaxes of the ellipse is $\tan\theta$, which is what we set out to prove.

The same formulas imply that at $\theta = 0$ the ellipse degenerates into a straight line, that is, $\theta = 0$ corresponds to a field linearly polarized in the xy plane. At $\theta = \pm\pi/4$ the ellipse is a circle, that is, $\theta = \pm\pi/4$ correspond to fields that are clockwise-polarized $(+)$ and counterclockwise-polarized $(-)$ in the xy plane.

Now let us establish the meaning of angle φ in (1.95). We put $2kz - 2\omega t = 0$. According to (1.99), the length of E is then equal to $E_\omega \sin\theta$, while the vector E, according to (1.92), (1.91), and (1.95), has the following form at $2kz - 2\omega t = 0$:

$$\begin{aligned} E &= E_\omega [\cos(\theta + \pi/4) - \cos(\theta - \pi/4)] \\ &\quad \times [\exp(i\varphi)\, e_+ - \exp(-i\varphi)\, e_-] \\ &= E_\omega \sin\theta \, [e_x \cos\varphi - e_y \sin\varphi]. \end{aligned} \tag{1.100}$$

We see that φ determines the orientation of the ellipse's semiaxes in relation to the laboratory Cartesian axes of coordinates, x and y. Figure 1.11 shows an ellipse described by the tip of the vector E with the passage of time.

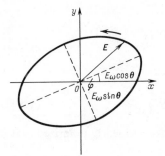

Figure 1.11. The elliptical path for the electric field strength E of the electromagnetic wave in elliptic polarization.

1.6.3. Propagation of an Elliptically Polarized Wave in a Linear Medium

As we saw in Section 1.1 in discussing the averaging of the angular momentum of an atom that is freely oriented in space over the various directions, the linear susceptibility tensor $\chi^{(1)}$ can be expressed in terms of the Kronecker symbol:

$$\chi_{ij}^{(1)} = \chi^{(1)}\delta_{ij}. \tag{1.101}$$

Hence, the linear polarization vector $\mathbf{P}^{(1)}$ at frequency ω is directed along the vector $\mathbf{E}_{\omega}'$ in the electromagnetic wave in the medium:

$$\mathbf{P}_{\omega}^{(1)} = \chi^{(1)}\mathbf{E}_{\omega}' \approx \chi^{(1)}\mathbf{E}_{\omega}. \tag{1.102}$$

We have written this relationship in the unperturbed-field approximation. Let us denote the difference between $\tilde{\mathbf{E}}'$ and $\tilde{\mathbf{E}}$ by $\delta\tilde{\mathbf{E}}$, that is, $\delta\tilde{\mathbf{E}} = \tilde{\mathbf{E}}' - \tilde{\mathbf{E}}$. This difference characterizes the polarization wave in the medium and obeys Maxwell's equation (1.44):

$$\nabla^2\delta\tilde{\mathbf{E}} + k^2\delta\tilde{\mathbf{E}} = -4\pi Nk^2\chi^{(1)}\tilde{\mathbf{E}}. \tag{1.103}$$

Assuming, following (1.92), that

$$\delta\tilde{\mathbf{E}} = \delta\mathbf{E}\exp(i\omega t - ikz), \tag{1.104}$$

substituting (1.104) into (1.103), and allowing for (1.92), we find that

$$\frac{\partial^2\delta\mathbf{E}}{\partial z^2} - 2ik\frac{\partial\delta\mathbf{E}}{\partial z} = -4\pi Nk^2\chi^{(1)}\mathbf{E}_{\omega}. \tag{1.105}$$

This yields

$$\delta\mathbf{E} = -2\pi ikzN\chi^{(1)}\mathbf{E}_{\omega}. \tag{1.106}$$

We can now find the total electric field strength in the medium:

$$\begin{aligned}
\tilde{\mathbf{E}}' &= \tilde{\mathbf{E}} + \delta\tilde{\mathbf{E}} \\
&= \left[1 - 2\pi ikzN\chi^{(1)}\right]\mathbf{E}_{\omega}\exp(i\omega t - ikz) \\
&\approx \mathbf{E}_{\omega}\exp(i\omega t - ik'z),
\end{aligned} \tag{1.107}$$

with

$$k' = k\left(1 + 2\pi N\chi^{(1)}\right) \tag{1.108}$$

(a similar result was obtained in Section 1.3 for a linearly polarized field).

We have therefore found that an elliptically polarized wave in a medium retains its polarization; that is, the ratio of the ellipse's semiaxes and the orientation of the semiaxes in relation to the laboratory coordinate system are conserved, while the wave vector changes: $\mathbf{k} \rightarrow \mathbf{k}'$. This statement remains valid when the linear susceptibility $\chi^{(1)}$ is a complex-valued quantity: the imaginary part of the susceptibility determines the exponential damping of the electromagnetic wave in the medium.

1.6.4. Conclusion

We conclude that a weak elliptically polarized electromagnetic wave propagates in a gaseous medium without changing its polarization.

1.7. CONCLUSION

In this chapter we have considered the main phenomena associated with linear optics in a medium in the form of an atomic gas, namely, the propagation of light in the medium, the reflection of light from an interface, the refraction of light at the interface, and the absorption of light by the medium. We found that these phenomena are described by well-known laws formulated on the macroscopic level. We established the relationship between these laws and the microscopic characteristic of the medium in all cases. Actually, all macroscopic laws of linear optics are determined by a single microscopic characteristic, the linear susceptibility of an atom, $\chi^{(1)}$. The magnitude of $\chi^{(1)}$ depends essentially on whether there are single-photon resonances between the light frequency and the frequency of atomic transitions. The most important property of the nonresonance linear susceptibility is that this quantity does not depend on the field strength. In contrast to the nonresonance linear susceptibility, the resonance susceptibility is determined by the resonance widths and thus can depend on the electric field strength in the electromagnetic wave. The problem becomes essentially nonlinear, and an expansion in a Taylor series in powers of the field strength becomes meaningless.

When we go over to the optical characteristics of a medium averaged over an (extremely) large number of atoms, we must distinguish between the nonresonance and resonance cases. In the absence of a resonance these characteristics are determined by a single macroscopic parameter, the number N of atoms per unit volume, while in the resonance case the averaged characteristics are determined, in addition to N, by the resonance width, which depends both on the field strength and on the macroscopic parameters of the medium, such as temperature and pressure.

By solving Maxwell's equations for light propagation in a linear medium characterized by a linear susceptibility, we arrive at the well-known phenomenological laws of linear optics.

Strictly speaking, the above description of linear optics is valid only for atomic gases within the framework of the model in which the light is assumed to be monochromatic and the gas consists of independent atoms. If we go outside these limits (nonmonochromatic light and high gas pressures) and, more than that, take a molecular gas or a transparent isotropic condensed substance (liquid, crystal, or glass) for the medium, qualitatively the above conclusions remain valid. New effects appear when light travels through anisotropic media.

2

Nonlinear Interaction of Light and Atoms

The nonlinear-optics effects that emerge when intense light interacts with atomic gases arise from the various nonlinear effects of light interacting with an isolated atom. The most important of these manifests itself in the nonlinearity of the atomic susceptibility. Here we are speaking of electronic susceptibility, which is caused by the various transitions of an optical electron between different discrete and continuous one-electron states in the atomic spectrum.

The linear atomic susceptibility, discussed in Chapter 1, constitutes only the first term in the expansion of the atomic polarization (that is, the mean dipole moment of the atom) in powers of the strength **E** of the electromagnetic field. Of course, there are always the higher-order terms in such an expansion. The question is: what is their relative role and dependence on the electric field strength of the light wave? In a weak field this role is negligible, with the result that the interaction of light with an atom is determined by the first term in the expansion, which is linear in the field strength (see Chapter 1). As the field strength is increased, the role of higher-order terms increases rapidly in view of their dependence on the field-strength amplitude as a power function.

Just as linear susceptibility conditions linear scattering of light in a gas (Chapter 1), nonlinear susceptibility conditions nonlinear light scattering, which will be discussed in this chapter. In Chapter 1 we found that the frequency of the scattered light may coincide with the frequency of the incident light (Rayleigh scattering), but may also differ by the frequency of an atomic transition (spontaneous Raman scattering). In nonlinear scattering the frequency spectrum of the scattered light proves to be much broader, since it contains harmonics of the incident light (due to multiphoton absorption) or sums or differences of frequencies in the case of several external fields.

As with linear susceptibility, nonlinear susceptibility and nonlinear light scattering are conveniently described within the framework of quantum mechanics as the absorption and emission of photons of various energies. While linear susceptibility is the result of the absorption by an atom of a single photon from the external light field, nonlinear susceptibility represents a more complex process, in which several photons from the external light field are absorbed or emitted (the photons can belong to different fields if there are several such fields). Nonresonance susceptibility is described in terms of virtual electron transitions, and resonance susceptibility in terms of real transitions. Ionization of atoms (including the nonlinear ionization, which we discuss in Section 2.4) constitutes a process that competes with the scattering of light by an atom, since under ionization the atom ceases to exist as a whole and transforms into an ion and a free electron (or several electrons).

The atomic polarization vector $\mathbf{P}$ can be expanded in powers of the electric field strength $\mathbf{E}$ in the light field

$$\mathbf{P} = \chi^{(1)}\mathbf{E} + \chi^{(2)}\mathbf{E}^2 + \chi^{(3)}\mathbf{E}^3 + \cdots, \tag{2.1}$$

so long as no resonances are present, the field strength being in the resonance denominators. This expansion, however, presupposes that $\mathbf{E}$ is small, that is, E is always much less than E_{at}, the atomic field strength. Higher-order perturbation theory can then be employed to describe polarization. Inasmuch as $\mathbf{P} = \langle \Psi | \mathbf{r} | \Psi \rangle$, where Ψ is the perturbed wave function of the atomic electron in the light field, we must determine Ψ within the framework of higher-order perturbation theory, after which we must take the dipole matrix element in the chosen order of perturbation theory. For instance, in third-order perturbation theory,

$$\mathbf{P}^{(3)} = \langle \Psi^{(0)} | \mathbf{r} | \Psi^{(3)} \rangle + \langle \Psi^{(1)} | \mathbf{r} | \Psi^{(2)} \rangle$$
$$+ \langle \Psi^{(2)} | \mathbf{r} | \Psi^{(1)} \rangle + \langle \Psi^{(3)} | \mathbf{r} | \Psi^{(0)} \rangle. \tag{2.2}$$

Here

$$\Psi = \Psi^{(0)} + \Psi^{(1)} + \Psi^{(2)} + \Psi^{(3)} + \cdots, \tag{2.3}$$

where $\Psi^{(0)}$ is the unperturbed atomic wave function, and $\Psi^{(K)} \propto E^K$ is the Kth term in the time-dependent perturbation expansion for the wave function.

At first glance, it would seem to follow from Eq. (2.1) that in describing nonlinear susceptibility and other nonlinear effects (e.g., nonlinear ionization) we can always ignore processes that have a higher order in the number of absorbed photons in comparison with lower-order processes. Actually, however, the situation is not so simple, because one must compare nonresonance, quasiresonance, and resonance processes. Often higher-order resonance processes have a higher probability than nonresonance lower-order processes. Hence, in each concrete case one must usually perform a quantitative analysis of the probabilities of various processes of nonlinear scattering and nonlinear absorption of light in order to determine the dominant phenomenon.

It goes without saying that the model of isolated atoms describes a real atomic gas fairly well if the latter has a low density. The criterion for this is the absence of collisions when there is no field and the absence of an effect produced by a strong light field on collisions between the particles constituting the gas. In a sufficiently strong field the role of collisions is not confined to the traditional collision broadening of the atomic levels. In this case, for the elementary quantum system we must take two atoms that interact during a collision together with the field of the electromagnetic wave acting on both atoms. Such collisions are known as *radiative*. In them, the collision widths are complicated functions of the strength of the light field. For this reason the interpretation of broadening effects poses certain difficulties (Yakovlenko, 1984).

Moreover, from the practical point of view, one must bear in mind that as the density of the atomic gas grows, the collisions of free electrons (arising, e.g., from the ionization of atoms by an electromagnetic field) with neutral atoms begin to play a role comparable with atom-atom collisions. When an electron is in a strong light field and collides with atoms, it may absorb photons from the light wave (inverse bremsstrahlung), so that its energy grows. The accelerated electrons ionize atoms, an electron avalanche develops, and a partially ionized plasma forms that is opaque to the incident electromagnetic radiation; because of this the gas gets heated and optical breakdown occurs. It is this breakdown that usually limits the increase in the atomic gas pressure for the purpose of increasing the nonlinear susceptibility of the medium.

In this chapter we consider the basic nonlinear phenomena that occur as a result of the interaction of a strong monochromatic light field with an isolated atom: in particular, the nonlinear susceptibilities, nonlinear scattering, Stark shifts of the atomic levels proportional to the square of the light intensity, multiphoton excitation, and nonlinear ionization.

In those cases where the light field acts not on an atom but on a molecule the phenomena are roughly the same. However, the more complex structure of molecules leads to a number of other phenomena, which have no analogy with the atomic case. Section 2.5 deals with such phenomena.

2.1. NONLINEAR ATOMIC SUSCEPTIBILITIES

2.1.1. Introduction

Let us start with the initial expression for the polarization, (2.1). In a weak electromagnetic field, the polarization **P** (per atom) is linear in the electric field strength $\mathbf{E}$ in the light wave. Now we will assume that $\mathbf{E}$ is so high that higher-order terms in the expansion of the polarization vector $\mathbf{P}$ in powers of $\mathbf{E}$ become important. If we ignore the fact that $\chi^{(K)}$ is a tensor (see below), then we can rewrite Eq. (2.1) at time t (at least symbolically) as

$$\mathbf{P} = \mathbf{P}^{(1)} + \mathbf{P}^{(2)} + \mathbf{P}^{(3)} + \cdots$$
$$= \chi^{(1)}\mathbf{E}(t) + \chi^{(2)}\mathbf{E}^2(t) + \chi^{(3)}\mathbf{E}^3(t) + \cdots. \tag{2.4}$$

The first term on the right-hand side is the linear part of atomic polarization (see Chapter 1), while the second term is known as the quadratic polarization, $\mathbf{P}^{(2)}$, the third the cubic polarization, $\mathbf{P}^{(3)}$, and so on. The $\chi^{(K)}$ are known as the nonlinear susceptibilities of the atom (of order K); they are tensors of rank $K + 1$. We will see later on that the nonlinear susceptibilities determine the nonlinear scattering and absorption of electromagnetic radiation in a medium, just as linear susceptibility $\chi^{(1)}$ determines the linear scattering and absorption of light (Chapter 1).

No general formula can be written for $\chi^{(K)}$, since different physical processes are determined by different terms in the Taylor series (2.4) and, as we will see later, there will be many such terms, depending on the sequence of absorption and emission of photons from the electromagnetic field and the ratio of the number of absorbed photons to the number of emitted. When the atom is illuminated by several light waves, the number of such terms will increase still further.

The quadratic nonlinear susceptibility $\chi^{(2)}$ is nonzero only in media without inversion centers, for instance, when the atoms or molecules of the media possess constant dipole moments or when certain crystals are chosen for the media. In atomic gases consisting of atoms in the ground state, we have $\chi^{(2)} = 0$ due to the parity conservation for atomic states, since the potential $V = \mathbf{r} \cdot \mathbf{E}$ of the dipole interaction of an atom with the electromagnetic field of the radiation changes the parity of an atomic state, while the susceptibility $\chi^{(2)}$, which appears in the third term of the perturbation expansion in powers of the interaction V, is proportional to $r_{nm}r_{mp}r_{pn}$, where n is the state of the atom in which the susceptibility is being determined. For $\chi^{(2)}$ to be nonzero, the state m of the atom must have a parity opposite to that of state n. Similarly, it follows from the dipole matrix element r_{mp} that the parity of state p must coincide with the parity of state n, both being opposite to the parity of state m. But then the dipole matrix element r_{pn} must be equal to zero, since this transition does not involve a change in parity. An obvious exception is a gas consisting of complex atoms in highly excited states, which are necessarily hydrogenlike; such atoms, as is well known (see Landau and Lifshitz, 1977, §77), possess a constant dipole moment at high orbital quantum numbers.

In media with inversion centers, such as atomic gases, the lowest-order nonzero nonlinear susceptibility is the cubic, $\chi^{(3)}$; the next nonlinear susceptibility is $\chi^{(5)}$, and so on.

Now we wish to investigate the tensor structure of the nonlinear susceptibility. In accordance with the expansion (2.4), the cubic susceptibility $\chi^{(3)}$ is a tensor of rank 4, or $\chi^{(3)} = \chi^{(3)}_{ijkl}$. Each i, j, k and l runs through three values, which correspond to the three Cartesian components (x, y, and z) or components in another orthogonal system of coordinates. The symbolic form of (2.4) can be elaborated, say, for the cubic polarization $\mathbf{P}^{(3)}(t)$ as

$$P_i^{(3)}(t) = \sum_{jkl} \int_0^\infty \int_0^\infty \int_0^\infty \chi^{(3)}_{ijkl}(\tau, \tau', \tau'') E_j(t - \tau'')$$

$$\times E_k(t - \tau') E_l(t - \tau) \, d\tau \, d\tau' \, d\tau'' \qquad (2.5)$$

Let us isolate the explicit frequency dependences of the physical quantities involved. As shown by (2.4), in a varying electromagnetic field with an electric field strength $\mathbf{E}(t)$, the polarization $\mathbf{P}$ is also a function of time t. Expansion of $\mathbf{P}$ in a Fourier series in t leads to the notion of polarization $\mathbf{P}_\nu$ at frequency ν. While in the linear case, obviously, $\nu = \omega$, in nonlinear processes ν does not necessarily coincide with the frequency of the electromagnetic field $\mathbf{E}(t) = 2\mathbf{E}_\omega\cos \omega t$, as we will see below. By the very definition of a Fourier expansion we have

$$\mathbf{P}(t) = \sum_\nu \mathbf{P}_\nu \exp(i\nu t). \qquad (2.6)$$

Note that it is $\mathbf{P}_\nu$ that is related to the experimentally observed characteristics when the interaction between radiation and gaseous medium is nonlinear, just as the linear polarization $\mathbf{P}^{(1)}$ determines the observed quantity when this interaction is linear.

As noted in the introduction to this chapter, we assume $\mathbf{E}(t)$ to be monochromatic. If, however, the field contains several frequency components, we must expand $\mathbf{E}(t)$ in a Fourier series in t and take only one Fourier component for each coefficient in (2.5). Then we can rewrite (2.5) for the Fourier components as

$$P_{i\nu}^{(3)} = \sum_{jkl} \chi^{(3)}_{ijkl} E_j(\omega_1) E_k(\omega_2) E_l(\omega_3). \qquad (2.7)$$

Here the dependence on time vanishes, and the frequency ν at which the polarization of the medium occurs is $\omega_1 + \omega_2 + \omega_3$. Of course, ω_1, ω_2, and ω_3 may be positive or negative, which in terms of photons means absorption or emission, respectively. Accordingly, the nonlinear susceptibility is denoted by

$$\chi_{ijkl}^{(3)} = \chi_{ijkl}^{(3)}(\nu; \omega_1, \omega_2, \omega_3) \text{ or } \chi_{ijkl}^{(3)}(-\nu; \omega_1, \omega_2, \omega_3). \tag{2.8}$$

As noted above, from the quantum-mechanical viewpoint, $\mathbf{P}$ constitutes the diagonal matrix element of the dipole moment of an atom with respect to the quantum-mechanical atomic state n considered here. Accordingly, the nonlinear susceptibility, say (2.8), is the diagonal matrix element with respect to n. This corresponds to nonlinear scattering of light by the atom in which the atom after being deflected returns to its initial state n. But if the nonlinear scattering process ends with the atom in another state p which differs from initial state n, it is obviously described by the off-diagonal matrix element of the dipole moment of the atom with respect to the initial (n) and final (p) states. If necessary, we will include in $\chi^{(K)}$ subscripts that denote the initial and final states; $\chi_{np}^{(3)}$, for instance.

Just as we expanded $\mathbf{E}$ in a Fourier series in t, we can expand the same quantity in the coordinates. Isolating the wave with a given wave vector $\mathbf{k}$ (note that the magnitude of $\mathbf{k}$ is related to the wave frequency thus: $k = \omega/c$) and using (2.7), we get

$$P_{i\nu}^{(3)}(\mathbf{k}) = \sum_{jkl} \chi_{ijkl}^{(3)}(\nu; \omega_1, \omega_2, \omega_3) E_j(\omega_1, \mathbf{k}_1)$$

$$\times E_k(\omega_2, \mathbf{k}_2) E_l(\omega_3, \mathbf{k}_3). \tag{2.9}$$

Here the Fourier expansion in $\mathbf{r}$ has been defined as follows:

$$\mathbf{E}(\omega_i, \mathbf{r}) = \sum_{\mathbf{k}_i} \mathbf{E}(\omega_i, \mathbf{k}_i) \exp(-i\mathbf{k}_i \cdot \mathbf{r}). \tag{2.10}$$

If we substitute this relationship into the right-hand side of (2.7), we find that the latter depends on $\mathbf{r}$ in the form $\exp(-i\mathbf{k}_1 \cdot \mathbf{r} - i\mathbf{k}_2 \cdot \mathbf{r} - i\mathbf{k}_3 \cdot \mathbf{r})$. If we perform a similar Fourier expansion in coordinates of the cubic polarization, that is, if we write the left-hand side of (2.7) as

$$\mathbf{P}_\nu^{(3)}(\mathbf{r}) = \sum_{\mathbf{k}} \mathbf{P}_\nu^{(3)}(\mathbf{k}) \exp(-i\mathbf{k} \cdot \mathbf{r}), \tag{2.11}$$

then by comparing the left- and right-hand sides of (2.7) we find the relationship between the wave vectors:

$$\mathbf{k} = \mathbf{k}_1 + \mathbf{k}_2 + \mathbf{k}_3. \tag{2.12}$$

This relationship reflects the law of momentum conservation in nonlinear light scattering.

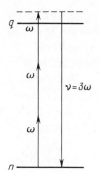

Figure 2.1. Emission-absorption diagram for third-harmonic excitation.

2.1.2. The Diagrammatic Technique for Nonlinear Atomic Susceptibility

The diagrammatic technique for describing the interaction of a monochromatic field in electromagnetic radiation with an atom has been studied in detail in Delone and Kraĭnov, 1985, and therefore is not described here. In Section 1.1.2 we have already dealt with the diagrammatic technique when the Feynman diagrams were associated with linear atomic susceptibilities. Let us now turn to nonlinear atomic susceptibilities.

Take the example of the cubic susceptibility (2.8) with $\omega_1 = \omega_2 = \omega_3 = \omega$ and $\nu = 3\omega$, that is,

$$\chi_{ijkl}^{(3)} = \chi_{ijkl}^{(3)}(3\omega;\ \omega, \omega, \omega). \tag{2.13}$$

Figure 2.1 shows the absorption and emission of photons by atomic states. In accordance with the rules for constructing Feynman diagrams, $\chi^{(3)}$ is depicted by the diagram in Figure 2.2, where the dashed lines designate the acts of absorption of the photons from the electromagnetic radiation of frequency ω. The wavy line depicts spontaneous photon emission (frequency $\nu = 3\omega$). Circles correspond to dipole matrix elements (e.g., $\mathbf{r}_{mp}$). Finally, vertical lines connecting the circles correspond to propagators, where the

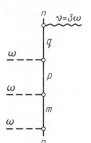

Figure 2.2. A Feynman diagram for the nonlinear atomic susceptibility describing the process of third-harmonic excitation depicted in Figure 2.1.

energy of the initial atomic state n is subtracted from the state with given quantum numbers (e.g., p), and so is the energy of all the photons absorbed in the transition from state n to the given state. Correspondingly, the energy of emitted photons must be added. For instance, the denominator in the propagator corresponding to the vertical line denoted in Figure 2.2 by p is $\omega_{pn} - 2\omega$. Summation of all the intermediate atomic states is tacitly assumed (in our case, summation is over states m, p, and q). Both figures clearly show that the process of nonlinear interaction of light with an atom incorporates multiphoton transitions of the atomic electron.

The Feynman diagram in Figure 2.2 determines, by its form, the diagonal matrix element of the dipole moment operator with respect to state n. Off-diagonal matrix elements are determined similarly (the lower and upper states in the Feynman diagrams representing these matrix elements differ). Such diagrams will be needed, for instance, in the resonance case, where the diagonal diagram of the higher-order nonlinear susceptibility can be represented by a product of lower-order off-diagonal matrix elements. They will also be needed when we consider multiphoton excitation and nonlinear ionization.

It follows that the diagram in Figure 2.2 for the cubic susceptibility is represented as follows:

$$\chi^{(3)}_{ijkl}(3\omega; \omega, \omega, \omega) = \sum_{mpq} r^i_{nm} r^j_{mp} r^k_{pq} r^l_{qn}$$

$$\times \left[(\omega_{mn} - \omega)(\omega_{pn} - 2\omega)(\omega_{qn} - 3\omega) \right]^{-1} + \cdots .$$

(2.14)

In Chapter 3 we will see that this diagram determines the physical process of third-harmonic excitation in the medium. As (2.14) shows, it corresponds

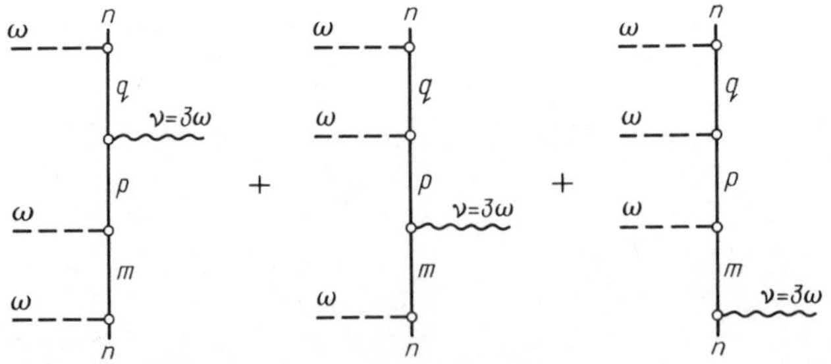

Figure 2.3. Other Feynman diagrams for the nonlinear atomic susceptibility describing third-harmonic excitation.

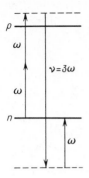

Figure 2.4. Emission-absorption diagram for third-harmonic excitation described by the first diagram shown in Figure 2.3.

to fourth-order terms in the quantum-mechanical perturbation theory of the interaction of the electromagnetic field with an atom.

Of course, in addition to the diagram depicted in Figure 2.2 there are similar diagrams that contribute to the cubic susceptibility (2.13). These are shown in Figure 2.3. They all differ in the order in which photons are absorbed and emitted. Each obeys the law of energy conservation: $\nu = \omega + \omega + \omega = 3\omega$. Together with (2.14) they determine the complete expression for the cubic susceptibility:

$$\chi^{(3)}_{ijkl}(3\omega; \omega, \omega, \omega) = \sum_{mpq} r^i_{nm} r^j_{mp} r^k_{pq} r^l_{qn}$$

$$\times \left\{ \left[(\omega_{mn} - \omega)(\omega_{pn} - 2\omega)(\omega_{qn} - 3\omega) \right]^{-1} \right.$$

$$+ \left[(\omega_{mn} - \omega)(\omega_{pn} - 2\omega)(\omega_{qn} + \omega) \right]^{-1}$$

$$+ \left[(\omega_{mn} - \omega)(\omega_{pn} + 2\omega)(\omega_{qn} + \omega) \right]^{-1}$$

$$\left. + \left[(\omega_{mn} + 3\omega)(\omega_{pn} + 2\omega)(\omega_{qn} + \omega) \right]^{-1} \right\}. \quad (2.15)$$

Figure 2.4 illustrates the process of absorption and emission of photons corresponding to the first Feynman diagram in Figure 2.3.

The Feynman diagrams for other nonlinear susceptibilities can be constructed in a similar manner. They differ from the one just discussed in the ratio of the number of absorbed photons to that of emitted photons, as well as in the frequency of cubic polarization.

2.1.3.   Examples of Cubic Susceptibilities

Let us take another cubic susceptibility, the one resulting from the action of a single electromagnetic field of frequency ω (just as in the previous case):

$$\chi^{(3)}_{ijkl} = \chi^{(3)}_{ijkl}(\omega; \omega, -\omega, \omega), \quad (2.16)$$

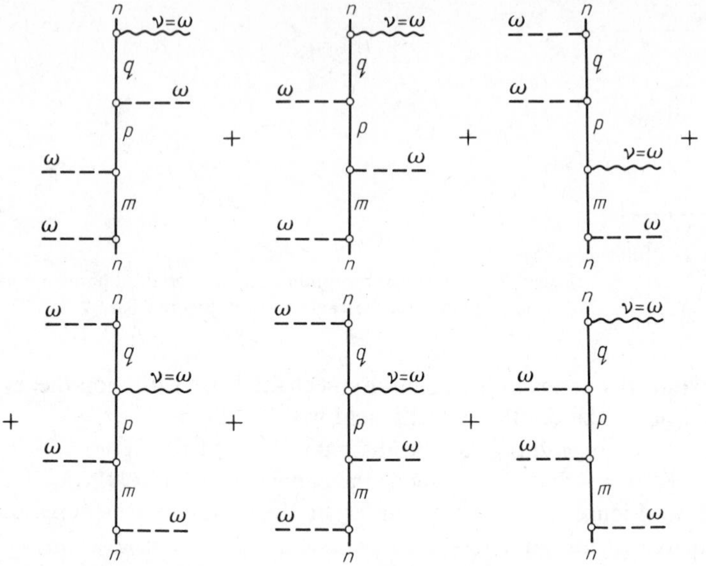

Figure 2.5. Feynman diagrams for the cubic susceptibility at frequency ω of the incident electromagnetic wave.

but differing from the one discussed above in the ratio of the number of absorbed photons to that of emitted photons. It is described by the Feynman diagrams depicted in Figure 2.5. In Figure 2.6 we demonstrate, by way of example, the process of absorption and emission of photons corresponding to the first diagram in Figure 2.5.

The analytical expression corresponding to these diagrams has the following form (according to the rule given above for the interpretation of such diagrams):

$$
\chi_{ijkl}^{(3)}(\omega; \omega, -\omega, \omega) = \sum_{mpq} r_{nm}^i r_{mp}^j r_{pq}^k r_{qn}^l
$$
$$
\times \Big\{ \left[(\omega_{mn} - \omega)(\omega_{pn} - 2\omega)(\omega_{qn} - \omega) \right]^{-1}
$$
$$
+ \left[(\omega_{mn} - \omega)\omega_{pn}(\omega_{qn} - \omega) \right]^{-1}
$$
$$
+ \left[(\omega_{mn} - \omega)\omega_{pn}(\omega_{qn} + \omega) \right]^{-1}
$$
$$
+ \left[(\omega_{mn} + \omega)(\omega_{pn} + 2\omega)(\omega_{qn} + \omega) \right]^{-1}
$$
$$
+ \left[(\omega_{mn} + \omega)\omega_{pn}(\omega_{qn} + \omega) \right]^{-1}
$$
$$
+ \left[(\omega_{mn} + \omega)\omega_{pn}(\omega_{qn} - \omega) \right]^{-1} \Big\}. \qquad (2.17)
$$

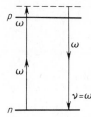

Figure 2.6. Emission-absorption diagram for nonlinear excitation of a polarization wave with frequency ω coinciding with the frequency of the incident electromagnetic wave. This corresponds to the first Feynman diagram in Figure 2.5.

Below [Eq. (3.53)] we will see that this cubic susceptibility determines the first nonlinear correction term to the refractive index of the atomic gas or to the dielectric constant ϵ_ω of this gas. In particular, if we direct the z axis along the vector $\mathbf{E}_\omega$, then

$$\epsilon_\omega = 1 + 4\pi N\chi^{(1)}(\omega; \omega) + 4\pi N\chi^{(3)}_{iizz}(\omega; \omega, -\omega, \omega)\mathbf{E}_\omega^2, \quad (2.18)$$

where N is the number of atoms of the gas per unit volume.

In Chapter 1 we found that for a homogeneous isotropic medium, of which atomic gases are an example, $\chi^{(1)}_{ij}$ is diagonal in components i, j, that is, $\chi^{(1)}_{ij} = \chi^{(1)}\delta_{ij}$. Hence, in linear optics the dielectric constant of the medium is the same in all directions with respect to the vector $\mathbf{E}_\omega$ in the radiation field.

In nonlinear optics, on the contrary, a strong light field results in anisotropy of the medium. If an infinitely weak field of frequency ν is directed in such a manner that its strength $\mathbf{E}_\nu$ is parallel to $\mathbf{E}_\omega$, then in accordance with (2.18) the dielectric constant of the gaseous medium for this field, denoted by $\epsilon_\nu^\parallel$, has the form

$$\epsilon_\nu^\parallel = 1 + 4\pi N\chi^{(1)}(\nu; \nu) + 4\pi N\chi^{(3)}_{zzzz}(\nu; \omega, -\omega, \nu)\mathbf{E}_\omega^2. \quad (2.19)$$

But if we direct $\mathbf{E}_\nu$ along the x axis, which is perpendicular to the z axis, then for the dielectric constant we get [see (2.18)]

$$\epsilon_\nu^\perp = 1 + 4\pi N\chi^{(1)}(\nu; \nu) + 4\pi N\chi^{(3)}_{xxzz}(\nu; \omega, -\omega, \nu)\mathbf{E}_\omega^2. \quad (2.20)$$

Subtracting (2.20) from (2.19), we find that

$$\epsilon_\nu^\parallel - \epsilon_\nu^\perp = 4\pi N\left(\chi^{(3)}_{zzzz} - \chi^{(3)}_{xxzz}\right)\mathbf{E}_\omega^2. \quad (2.21)$$

Thus, the difference between the dielectric constants in the parallel and perpendicular directions (in relation to $\mathbf{E}_\omega$) is nonzero only due to the nonlinear correction term in the dielectric constant. This phenomenon is known as the electron Kerr effect in atoms.

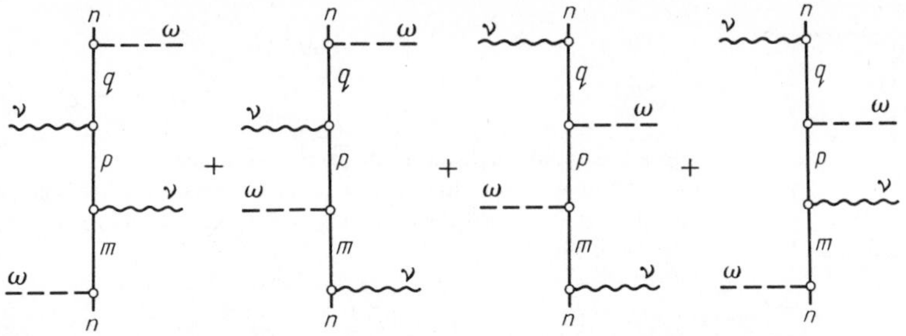

Figure 2.7. Feynman diagrams for cubic susceptibility describing stimulated Raman scattering.

The nonlinear correction term in (2.18) is the cause of a well-known nonlinear-optics phenomenon: the self-focusing of intense light beams in gases (Chapter 3), related to the inhomogeneity in the magnitude of the electric field in the radiation wave in the direction normal to the one in which the wave propagates.

The nonlinear cubic susceptibility $\chi^{(3)}_{ijkl}(\nu; \omega, \nu, -\omega)$ is also responsible, as we will see in Chapter 3, for the widely known process of stimulated Raman scattering of light. The complete expression for this susceptibility can be written if we employ the rules for constructing Feynman diagrams discussed above. In Figure 2.7 we show the Feynman diagrams for the nonlinear susceptibility that determines stimulated Raman scattering, while in Figure 2.8 we show the atomic levels and absorption and emission of photons during stimulated Raman scattering, which clearly shows the essence of the phenomenon. We will consider stimulated Raman scattering, which uses the above-noted nonlinear susceptibility, in greater detail in Chapter 3 (see also Section 2.2).

Obviously there are other cubic susceptibilities, having different ratios of the number of absorbed photons to that of emitted photons and different frequencies of the interacting fields. For instance, in the case of three light waves with different frequencies ω_1, ω_2, and ω_3, there appears the cubic susceptibility of the form $\chi^{(3)}_{ijkl}(\nu; \omega_1, \omega_2, \omega_3)$, with $\nu = \omega_1 + \omega_2 + \omega_3$ the frequency of the induced nonlinear-polarization wave. The fact that the nonlinear susceptibility depends on the frequencies of all three incident

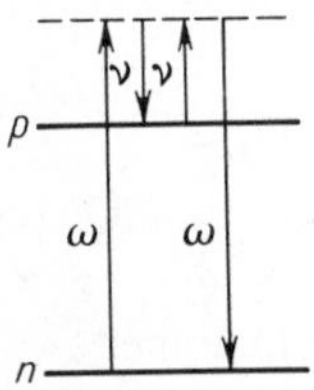

Figure 2.8. Emission-absorption diagram for stimulated Raman scattering.

waves is proof that on the atomic level there are coupled waves in a nonlinear medium (see Section 3.4).

2.1.4. Higher-Order Nonlinear Susceptibilities

As the intensity of the light interacting with an atom is increased, higher-order terms in the expansion of the atomic polarization in powers of the electric field strength, (2.1), become more and more important. Following the cubic susceptibility, there is the nonlinear susceptibility $\chi^{(5)}$ that plays a significant role. The simplest physical phenomenon caused by $\chi^{(5)}$ is the excitation of the fifth harmonic of the field of the radiation incident on the medium. Other processes are also possible.

The excitation of higher harmonics in the interaction of radiation of the optical frequency range with atomic gases is one of the most promising methods for creating sources of coherent radiation in the ultraviolet frequency range (Section 3.2); whence the attention paid to higher-order nonlinear susceptibilities.

At present no problems arise from the necessity of using high fields in the optical range in order to investigate the effects introduced by higher-order nonlinear susceptibilities. A pulsed laser easily generates electric field strengths up to the atomic strength. The basic difficulty that appears in the study of higher-order nonlinear susceptibilities lies in the need to distinguish between effects brought on by *direct nonlinear processes* caused by higher-order nonlinear susceptibilities, and those brought on by *cascade nonlinear processes*, caused by lower-order nonlinear susceptibilities. These processes proceed through occupied excited states and are caused by new electromagnetic waves appearing in the medium. For instance, the excitation of the fifth harmonic, already mentioned, which is determined by the nonlinear susceptibility $\chi^{(5)}(5\omega; \omega, \omega, \omega, \omega, \omega)$, may be in competition with the process caused by cascade interaction, in which the third harmonic (frequency 3ω) of the incident radiation is excited and interacts with two photons from the initial electromagnetic wave of frequency ω. The corresponding (cubic) nonlinear susceptibility is $\chi^{(3)}(5\omega; 3\omega, \omega, \omega)$. Calculations and experiments (Doitcheva, Mitev, Pavlov, and Stamenov, 1978) have shown that the susceptibilities for the cascade nonlinear processes are only slightly smaller than those for the direct processes with the same number of absorbed photons. One must therefore include cascade processes in the picture.

At present, the higher-order susceptibilities of atomic gases are known up to $\chi^{(9)}$. A detailed discussion of calculations and measurements of higher-order susceptibilities is contained in Bloembergen, 1977; Geller and Popov, 1981; and Hanna, Yuratich, and Cotter, 1979.

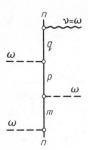

Figure 2.9. The Feynman diagram describing the secular term in the cubic susceptibility at frequency ω of the incident electromagnetic wave.

2.1.5. Removal of Secular Terms in Nonresonance Nonlinear Susceptibilities

In the analytical expressions for the nonlinear susceptibilities discussed above there are terms that formally become infinite even in the general nonresonance case. Consider, for example, the Feynman diagram for the cubic susceptibility $\chi_{ijkl}^{(3)}(\omega; \omega, -\omega, \omega)$, which is responsible for the nonlinear part in the index of refraction at frequency ω. This diagram is shown in Figure 2.9 (which is simply the second diagram in Figure 2.5). Here we consider solely the term in this diagram with $p = n$. The corresponding analytical expression [see the second term in (2.17)] is

$$\sum_{mq} r_{nm}^i r_{mn}^j r_{nq}^k r_{qn}^l \left[(\omega_{mn} - \omega) \omega_{nn} (\omega_{qn} - \omega) \right]^{-1}, \tag{2.22}$$

which is formally equal to infinity because $\omega_{nn} = 0$. Such divergent terms are called *secular*.

The secular terms disappear when, instead of using the energies $\mathscr{E}_n^{(0)}$ of the unperturbed Hamiltonian in determining the average value of the dipole moment of the atom in state n in the propagators in the terms of the (2.22) type, one employs the exact energy values $\tilde{\mathscr{E}}_n$ of the Hamiltonian that includes the interaction of the atom with the electromagnetic field. Then (2.22) contains $\tilde{\omega}_{nn} = \tilde{\mathscr{E}}_n - \mathscr{E}_n$, which is nonzero, and the infinities disappear. Moreover, it has been found that in the nonlinear susceptibilities—in particular, in the expression (2.17) for $\chi^{(3)}(\omega; \omega, -\omega, \omega)$—the terms containing $\tilde{\omega}_{nn}$ cancel out, so that the summation in (2.17) can be carried out over states $p \neq n$, for which, already within the fourth-order perturbation expansion considered here, the energy of state n in the propagators can be taken as the unperturbed energy (the perturbations of this energy by the field will be felt only in sixth-order terms).

Fourth-order terms, which contribute to $\chi^{(3)}(\omega; \omega, -\omega, \omega)$, will appear within the framework of the approach considered with exact energies $\tilde{\mathscr{E}}_n$, due to the expression for the linear susceptibility written in the form

$$\tilde{\chi}_{ij}^{(1)}(\omega;\,\omega) = \sum_m r_{nm}^i r_{mn}^j \left[\left(\tilde{\mathscr{E}}_m - \tilde{\mathscr{E}}_n - \omega \right)^{-1} + \left(\tilde{\mathscr{E}}_m - \tilde{\mathscr{E}}_n + \omega \right)^{-1} \right]. \quad (2.23)$$

Here we must bear in mind that

$$\tilde{\mathscr{E}}_n = \mathscr{E}_n - \chi_{ij}^{(1)} E_i E_j. \quad (2.24)$$

Substituting (2.24) into (2.23) and expanding (2.23) in a series in powers of $E_i E_j$, we obtain the terms that contribute to $\chi^{(3)}(\omega;\,\omega,\,-\omega,\,\omega)$. Detailed expressions for these terms can be found in Hanna, Yuratich, and Cotter, 1979, Section 2.6.5.

Note that there is no problem with secular terms in the linear susceptibility, whence no mention of them in Chapter 1. In the linear case the terms vanished for a far simpler reason: at $m = n$ the diagonal matrix element r_{nm} of the position coordinate vanishes because of parity conservation in the atomic states. But with nonlinear susceptibility this is not so, since states p and n in (2.17) have the same parity.

2.1.6. Calculations of Nonresonance Nonlinear Susceptibilities

These are carried out using the time-dependent higher-order perturbation theory in the interaction of field and atom (see Delone and Kraĭnov, 1985, Chapter 2). The difficulties are the same as in calculating the linear-optics constants (Section 1.1.6). The reader will recall that one difficulty is the need to allow for an infinite number of states in the continuous spectrum of the atoms, states over which the intermediate summation must be carried out. At the same time, calculation of the linear susceptibility requires only one summation, while in the nonlinear case there are several such summations, which complicates matters additionally (see Rapoport, Zon, and Manakov, 1978). Another difficulty is associated, as already noted in Section 1.1.6, with the fact that there are no rigorous analytical expressions for the wave function of an optical electron in a complex atom.

As always, the simplest calculation involves the nonlinear susceptibility of a hydrogen atom. Let us take the example of calculating $\chi^{(3)}(3\omega;\,\omega,\,\omega,\,\omega)$ as a function of ω. This calculation was carried out by Rapoport, Zon, and Manakov, 1978, where the Coulomb Green function was employed. The results for the ground state, $1S$, of hydrogen are depicted in Figure 2.10, where the resonances correspond to electromagnetic field frequencies $\omega = \omega_{qn}/3$, with $n = 1S$, and $q = 2P, 3P$ the excited states of the hydrogen atom. We see that $\chi^{(3)}$ changes sign as the frequency passes through the resonance values, which agrees with (2.15). The sign also changes in the interresonance interval between the $3P$ and $4P$ levels.

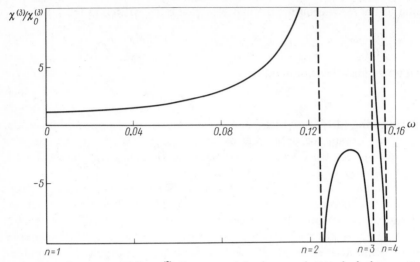

Figure 2.10. Cubic susceptibility $\chi^{(3)}_{zzz}(3\omega; \omega, \omega, \omega)$ for the ground state of a hydrogen atom as a function of frequency ω of the incident electromagnetic wave (in units of the dc value of this susceptibility, $\chi^{(3)}_0$, $\omega \to 0$).

The results of calculations of the nonlinear susceptibility of a hydrogen atom are only of illustrative value. Practically one needs the nonlinear susceptibilities of more complex atoms, such as alkali atoms, alkaline-earth atoms, and noble atoms (see also Section 3.2).

An example of calculations of the susceptibility $\chi^{(3)}(3\omega; \omega, \omega, \omega)$ for alkali atoms can be found in Miles and Harris, 1973. The calculation was carried out according to Eq. (2.15). The integration over angular variables in (2.15) and the writing of $\chi^{(3)}$ in the form of a combination of reduced matrix elements are carried out by the common methods of angular momentum theory and are treated in detail in Orr and Ward, 1971 (see also Sobelman, 1979). For an isotropic medium, an example of which is a gaseous medium, and an electromagnetic field linearly polarized along the z axis, $\chi^{(3)}_{ijkl}(3\omega; \omega, \omega, \omega)$ has only one nonzero component, $\chi^{(3)}_{zzzz}(3\omega; \omega, \omega, \omega)$. This is easily explained, since $\mathbf{r} \cdot \mathbf{E}_\omega = zE_\omega$. Numerical calculations for alkali atoms are reported in Miles and Harris, 1973. The radial dipole matrix elements $\mathbf{r}_{mn}$ in (2.15) were calculated in the hydrogen-like approximation. Twelve discrete atomic states were taken into account in the sums in (2.15).

Figure 2.11 gives the dependence of $\chi^{(3)}(3\omega; \omega, \omega, \omega)$ on the wavelength $\lambda = 2\pi c/\omega$ for the ground state of a sodium atom (Miles and Harris, 1973). One can clearly see that $\chi^{(3)}$ exhibits a multitude of resonances [in accord

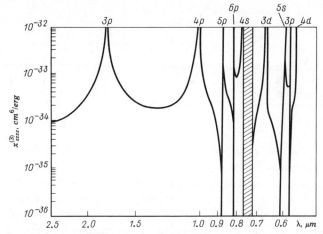

Figure 2.11. Cubic susceptibility $\chi^{(3)}_{zzzz}(3\omega; \omega, \omega, \omega)$ for the ground state of a sodium atom as a function of wavelength $\lambda = 2\pi c/\omega$. The hatched section corresponds to a region with a large number of closely spaced resonances.

with the analytical expression (2.15)]. Calculations have been carried out in the 0.5- to 3-μm wavelength range. Resonances in Figure 2.11 correspond to frequencies $\omega = \omega_{mn}$ and $\omega = \omega_{pn}/2$; the resonance frequencies $\omega = \omega_{qn}/3$ are not shown, because they correspond to too high wavelengths (or too low frequencies). We see that $\chi^{(3)}$ may vanish in the interresonance intervals, but this is not a necessary condition.

In Eicher, 1975, the calculation of $\chi^{(3)}(3\omega; \omega, \omega, \omega)$ was refined by allowing for spin-orbit splitting of the energy levels of the alkali atoms, which led to more precise values of the propagators in (2.15). Moreover, more correct values (than those given in Miles and Harris, 1973) of the dipole matrix elements were substituted into (2.15). Obviously, allowing for the spin-orbit splitting is necessary when the misfit between the frequency of the atomic transition and that of the field, ω (or the second harmonic 2ω), is comparable with this splitting. Because of the large magnitude of spin-orbit splitting in sodium atoms, this is important at practically all frequencies ω.

The calculation of $\chi^{(3)}(3\omega; \omega, \omega, \omega)$ for alkali atoms has also been carried out in Manakov, Ovsyannikov, and Rapoport, 1975 (see also Rapoport, Zon, and Manakov, 1978). Instead of direct summation over the intermediate virtual states of the atom, the authors used the Green-function method. The Green function was built into the model-potential approximation. Numerical calculations were carried out only for fixed values of the field frequency ω (specifically, for Nd-glass and ruby-laser frequencies).

Table 2.1   Nonlinear Atomic Susceptibilities $\chi^{(K)}(K\omega; \omega, \ldots)$

Unit	$\chi^{(K)}$			
	$K = 3$	5	7	9
cm³ (cm³/erg)$^{(K-1)/2}$	10^{-33}	10^{-43}	10^{-48}	10^{-58}
Atomic units	10^6	10^2	10	1

Similar calculations are reported in Manakov, Ovsyannikov, and Rapoport, 1975, for noble-gas atoms.

In making such calculations and comparing the results obtained with different atoms, one must bear in mind that the above formulas for the nonlinear susceptibilities were based on the assumption that the electric field strength in the electromagnetic monochromatic wave is $E(t) = E_\omega \exp(i\omega t) + E_\omega^* \exp(-i\omega t) = 2E_\omega \cos \omega t$. But if we put $E(t) = E_\omega \cos \omega t$, we must introduce the factor 2^{-K} into the perturbation expansions and hence into the above formulas.

Aside from the works cited above, numerous calculations of other nonlinear susceptibilities have been performed (e.g., Kielich, 1981, and Rapoport, Zon, and Manakov, 1978); these differ little from the calculations of $\chi^{(3)}(3\omega; \omega, \omega, \omega)$ discussed above.

So that the reader may get a clearer picture, we conclude this section by bringing together all the results of calculations performed by Doitcheva, Mitev, Pavlov, and Stamenov, 1978, for higher-order nonlinear susceptibilities of sodium atoms of the form $\chi^{(K)}(K\omega; \omega, \omega, \ldots, \omega)$. These data can be employed to estimate the order of magnitude of other nonlinear susceptibilities as well. Note that $\chi^{(K)}$ have different dimensionalities depending on K. The orders of magnitude of these quantities listed in Table 2.1 are valid for nonlinear susceptibilities in typical interresonance segments or in the static limit $\omega \to 0$. In atomic units ($\hbar = e = M_e = 1$) all $\chi^{(K)}$ are of the order of unity. (In Table 2.1 the results are given for the ground state of a sodium atom in the field of a Nd-glass laser; the atomic unit of $\chi^{(K)}$ is $a^3(a^3/2R_\infty)^{(K-1)/2}$, where a is the Bohr radius and R_∞ is the Rydberg constant.)

2.1.7. Conclusion

The above examination of nonlinear atomic susceptibilities has shown that they lead to a much richer spectrum of frequencies emitted by an atom than in the case where only the linear susceptibility is important. Nonlinear susceptibilities determine the induced dipole moment of the atom in a given state n, and this moment oscillates with time. But the frequencies of these oscillations, ν, may differ from the frequency ω of the incident electromagnetic wave (say, $\nu = 3\omega$). In accord with the semiclassical theory of dipole radiation, the atom will emit photons whose frequency is ν. This constitutes the main difference from the linear case (Section 1.1.6), where $\nu = \omega$.

Thus, for elastic nonlinear scattering, where the final atomic state coincides with the initial, the frequency of the emitted photons may differ from that of the incident radiation. Similarly, for inelastic nonlinear light scattering on an atom, where the final atomic state p differs from the initial state n, the spectrum of the emitted frequencies ν is much richer than the frequency $\omega - \omega_{pn}$ of the spontaneous Raman scattering in linear optics.

2.2. RESONANCE NONLINEAR ATOMIC SUSCEPTIBILITIES

2.2.1. Introduction

The expressions for nonlinear susceptibilities given in Section 2.2 demonstrate that all contain a certain number of propagators. If the denominators in these propagators are not anomalously small, the effects of the nonlinearity are usually small, provided that the electric field strength in the field of the incident electromagnetic radiation is low compared to the characteristic atomic field strength. (The reader will recall that in atomic units, $\chi^{(K)} \sim 1$.) Hence, to observe nonlinear effects, one must realize the quasiresonance situation, when the resonance misfit, being small compared with the field's frequency, is large compared with the characteristic widths of atomic levels; or the resonance situation, where the resonance misfit becomes smaller than the resonance width. One must bear in mind, however, that under exact resonance conditions there appears a competing channel by which an optical electron travels to the continuous spectrum. This transition may be either single-photon or multiphoton. Obviously, the competition of single-photon ionization is the most important. The probability of single-photon ionization from excited states was discussed earlier [see (1.31)]; the case of multiphoton ionization will be considered in Section 2.4.

In both the nonlinear and the linear case (see Section 1.2.2), any resonance can be described by formulas of the type (2.15), written within the scope of perturbation theory, by adding a quantity $i\Gamma_m$ to the denominator of the corresponding propagator, where Γ_m is the effective resonance width. Of course, quantitatively, the Lorentzian profile only holds for the broadening mechanism that is due to the presence of a natural lifetime of the resonance atomic state. Such mechanisms as the Doppler broadening have profiles that differ from the Lorentzian. As for the

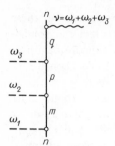

Figure 2.12. One of the Feynman diagrams for the nonlinear susceptibility $\chi^{(3)}(\nu; \omega_1, \omega_2, \omega_3)$ describing the excitation of a nonlinear polarization wave with the sum frequency $\nu = \omega_1 + \omega_2 + \omega_3$.

resonance widths, their physical origin is the same as in the linear case (Section 1.2.4). The difference is that, first, in the nonlinear case the resonance may be single-photon and multiphoton and, second, several resonances (rather than one) may be simultaneously realized in the nonlinear susceptibilities.

Let us see how to account for an intermediate resonance, using the nonlinear susceptibility $\chi^{(3)}(\nu; \omega_1, \omega_2, \omega_3)$ as an example. This susceptibility describes the process of excitation of the frequency $\nu = \omega_1 + \omega_2 + \omega_3$. One Feynman diagram for this quantity is shown in Figure 2.12, and in Figure 2.13 we give the schematic of the process of absorption and emission of photons for the given physical phenomenon. The following analytical expression can be associated with the abovementioned diagram, where we have allowed for the presence of effective widths Γ_m in the appropriate denominators:

$$\chi^{(3)}_{ijkl}(\nu; \omega_1, \omega_2, \omega_3) = \sum_{mpq} r^i_{nm} r^j_{mp} r^k_{pq} r^l_{qn}$$

$$\times \left[(\omega_{mn} - \omega_1 + i\Gamma_m)(\omega_{pn} - \omega_1 - \omega_2 + i\Gamma_p) \right.$$

$$\left. \times (\omega_{qn} - \nu + i\Gamma_q) \right]^{-1} + \cdots. \quad (2.25)$$

Here $\Gamma_m, \Gamma_p, \Gamma_q$ are the effective widths of the corresponding atomic states

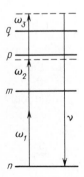

Figure 2.13. Emission-absorption diagram for the process described by the Feynman diagram in Figure 2.12.

(the reasoning for this broadening will not be discussed here). Of course, these widths must be added in the denominators only if the misfit of the corresponding intermediate resonance is of the order of the resonance width or less.

In a similar way the widths must be added to the many other terms left out in (2.25).

2.2.2. The Effect of Level Population on Nonlinear Susceptibility

Equation (2.25) for the resonance nonlinear susceptibility was written on the assumption that the intermediate atomic states m, p, and q are not, in reality, populated by the field of the electromagnetic wave. This is true only for large resonance misfits or in the case of sufficiently large widths of the resonance states, when the population of the initial state is close to unity and that of the other states is close to zero.

Let us now discuss the opposite case, when this is not so, and take as an example the same diagram in Figure 2.12 representing the absorption of photons with frequencies ω_1, ω_2, and ω_3. Suppose that in this process the intermediate level p is really populated by two-photon absorption of photons with frequencies ω_1 and ω_2. Thus, the population of state p, which we denote by N_p, differs substantially from zero and is comparable with that of the initial level n, denoted by N_n. Hence, now N_n differs from unity. The effect of population on the linear susceptibility was discussed in Chapter 1, where it was related to the one-photon resonance transition $n \rightarrow m$, while here we are dealing with the two-photon resonance transition $n \rightarrow p$. Let us determine the effect that population of level p has, under such a transition, on the nonlinear susceptibility $\chi^{(3)}$.

If an atom is in state n, its nonlinear susceptibility is determined by (2.25). The population probability for state n is N_n, so when an atom is in state n, its average dipole moment and hence $\chi^{(3)}$ are proportional to N_n. Thus, we must multiply (2.25) by N_n. But this does not solve the problem. Since there is considerable probability of the atom being in state p, its average dipole moment is $\langle n|\mathbf{r}|n \rangle N_n + \langle p|\mathbf{r}|p \rangle N_p$. To determine $\langle p|\mathbf{r}|p \rangle$ we must know the nonlinear susceptibility $\chi^{(3)}$ in state p. For the process depicted in Figure 2.12, the nonlinear susceptibility in state p is determined by the Feynman diagram shown in Figure 2.14. This diagram has corresponding to it the following analytical expression:

$$\chi^{(3)}_{ijkl}(p) = \sum_m r^i_{nm} r^j_{mp} (\omega_{mp} + \omega_2)^{-1} (\omega_{np} + \omega_1 + \omega_2 - i\Gamma_p)^{-1}$$
$$\times \sum_q r^k_{pq} r^l_{qn} (\omega_{qp} - \omega_3)^{-1}. \tag{2.26}$$

The sum over n is absent due to the resonance condition for the two-pho-

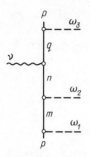

Figure 2.14. One of the Feynman diagrams for the nonlinear susceptibility in the resonance state p corresponding to the process depicted in Figure 2.13.

ton excitation process, $\omega_1 + \omega_2 \approx \omega_{pn}$. In addition, we have allowed for the width only in the resonance propagator. Similarly, due to the same condition, the sum over p disappears in (2.25), and we can rewrite (2.25) as

$$\chi_{ijkl}^{(3)}(n) = \left(\omega_{pn} - \omega_1 - \omega_2 + i\Gamma_p\right)^{-1}$$
$$\times \sum_m r_{nm}^i r_{mp}^j \left(\omega_{mn} - \omega_1\right)^{-1} \sum_q r_{pq}^k r_{qn}^l \left(\omega_{qn} - \nu\right)^{-1}. \quad (2.27)$$

Since $\omega_2 \approx \omega_{pn} - \omega_1$ because of the resonance condition, in the first propagator in (2.26) we can write $\omega_{mp} + \omega_2 \approx \omega_{mp} + \omega_{pn} - \omega_1 = \omega_{mn} - \omega_1$. Similarly, in the second propagator we write $\omega_{qp} - \omega_3 \approx \omega_{qn} - \omega_{pn} - \omega_3 = \omega_{qn} - \nu$. Now we see that the first sum in (2.26) coincides with the first sum in (2.27), and the second sum in (2.26) with the second sum in (2.27). As for the resonance propagators in (2.26) and (2.27), one can easily see that they are equal in magnitude and opposite in sign. Hence, we obtain $\chi_{ijkl}^{(3)}(n) = -\chi_{ijkl}^{(3)}(p)$. For the nonlinear susceptibility, in conditions where the resonance state p is really populated, we have the following expression:

$$\chi_{ijkl}^{(3)} = \left(N_n - N_p\right)\chi_{ijkl}^{(3)}(n), \quad (2.28)$$

.with $\chi_{ijkl}^{(3)}(n)$ determined by (2.27). For instance, if the populations of n and p become the same, the dipole moment of such a system vanishes, or, as is usually said, the atomic transition $n \to p$ becomes saturated.

Equation (2.27) can be written in a more abridged form:

$$\chi_{ijkl}^{(3)}(\nu; \omega_1, \omega_2, \omega_3) = \chi_{np}^{(1)}(-\omega_2; \omega_1)\chi_{pn}^{(1)}(\nu; \omega_3)$$
$$\times \left(\omega_{pn} - \omega_1 - \omega_2 + i\Gamma_p\right)^{-1}. \quad (2.29)$$

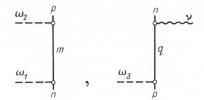

Figure 2.15. Feynman diagrams for the off-diagonal matrix element of the linear susceptibility that is present in the cubic susceptibility corresponding to the process depicted in Figure 2.13.

Here we have introduced the following notation:

$$\chi_{np}^{(1)}(-\omega_2; \omega_1) = \sum_m r_{nm}^i r_{mp}^j (\omega_{mn} - \omega_1)^{-1},$$
$$\chi_{pn}^{(1)}(\nu; \omega_3) = \sum_q r_{pq}^k r_{qn}^l (\omega_{qp} - \omega_3)^{-1}. \tag{2.30}$$

These two quantities are off-diagonal matrix elements of the linear susceptibility (see Chapter 1 and Section 2.1) and correspond to the atomic transition from state n to state p. They are represented by the Feynman diagrams depicted in Figure 2.15. Thus, in the resonance case, the nonlinear cubic susceptibilities are represented by the product of two linear (but off-diagonal) susceptibilities. This, obviously, simplifies the calculation of $\chi^{(3)}$ considerably.

2.2.3. The Physical Meaning of the Real and Imaginary Parts of Nonlinear Susceptibilities

In Section 1.2.2 we considered the resonance linear susceptibility and established that it possesses both a real and an imaginary part (in the nonresonance case, obviously, only the real part is nonzero). The real part is responsible for the linear part of the index of refraction at the frequency of the incident radiation. The imaginary part, on the other hand, is responsible for one-photon absorption of radiation, that is, for the reduction in intensity of the radiation passing through the medium. Radiation absorption, as we have seen, is accompanied by the transition of medium atoms into excited resonance states. Just as in the linear case, these states can be deexcited in a cascade manner, with the atom going over to a lower-lying state and, finally, to the ground state. The frequencies of the emitted photons differ from those of the incident photons, which means simply that there is disappearance of photons in the medium, or absorption of the incident light.

As for the real part of the nonlinear resonance susceptibility, it determines, as in the nonresonance case, the correction (quadratic in field strength) to the index of refraction of the medium [or the dielectric constant; see (2.18)–(2.20)]. In accordance with (2.29), the correction vanishes at exact resonance and changes sign in the neighborhood of the resonance.

We now turn to the imaginary part of the resonance nonlinear susceptibility. As Eq. (2.29) shows, this part, as in the linear case, is responsible for radiation absorption, but two-photon instead of one-photon. Here one absorbed photon has frequency ω_1 and the other ω_2.

If other resonances manifest themselves, the imaginary part of the nonlinear susceptibility may be responsible for one-photon absorption of the incident radiation accompanied by emission of a photon with another frequency. As we will see in Chapter 3, this occurs, for instance, in the process of stimulated Raman scattering. If we look at Eq. (2.29), a photon of frequency ω_1 from the incident radiation is absorbed by the medium, while a photon of the stimulated-Raman-scattering frequency $\omega_2 < 0$ is emitted at the same time. We will also see that the imaginary part of (2.29) determines the gain of the stimulated-Raman-scattering polarization wave at frequency $|\omega_2|$.

But if we turn to the process in which new polarization waves are formed in the medium, e.g., the process of third-harmonic excitation, when the intensity of such waves is small compared to the incident radiation intensity (in contrast with the process of propagation of such waves in a nonlinear medium, where they have a considerable intensity), the polarization vector proportional to the appropriate nonlinear susceptibility determines, via Maxwell's equations, the electric field strength in the polarization wave (see Chapter 3). Hence, the intensity of the emerging polarization wave is proportional to the square of the absolute value of the nonlinear susceptibility. For example, in Section 3.2 we will show that the intensity of the third harmonic, $I_{3\omega}$, is proportional to $|\chi^{(3)}(3\omega; \omega, \omega, \omega)|^2$ (and, of course, to the cube of the intensity I_ω of the incident radiation wave of frequency ω). In the resonance case this quantity is equal, obviously, to the sum of the squares of the real and imaginary parts of the resonance nonlinear susceptibility, while in the nonresonance case only the real part is nonzero. This statement is valid only if $I_{3\omega} \ll I_\omega$, where I_ω is the intensity of the incident radiation.

But when the intensity of the third-harmonic polarization wave, $I_{3\omega}$, becomes, after passage through the medium, so high that it can be compared to the intensity I_ω of the incident radiation (of frequency ω), we find ourselves in the former situation, where the real part Re $\chi^{(3)}$ determines the medium's index of refraction for the third harmonic (more precisely, its nonlinear part), while the imaginary part Im $\chi^{(3)}$ determines the (coefficient of) transformation of the radiation from the fundamental frequency to the third harmonic. The necessary relationships for various physical processes will be obtained in Chapter 3.

Up till now we have considered the real and imaginary parts of the resonance nonlinear susceptibility within the framework of the Taylor expansion in powers of the electric field strength [see (2.4)]. In particular, it was assumed that the difference in the populations, $N_n - N_p$, entering into (2.28) is a result of collisions and other mechanisms but not the presence of an external field. However, a strong light field diminishes this difference. A numerical calculation can be carried out within the framework of the resonance approximation for two levels n and p. In this case we have a two-photon resonance, since the transition from n to p is accompanied by the absorption of a photon of frequency ω_1 and a photon of frequency ω_2. Rautian, 1970, has performed a numerical calculation from which it follows that $N_n - N_p$ should be multiplied by $(1 + w_{pn}/\Gamma_p)^{-1}$, where w_{pn} is the probability of a two-photon transition $n \to p$ occurring under the action of a field of frequency ω_1 and a field of frequency ω_2. If, for the sake of simplicity, we assume that $\omega_1 = \omega_2 = \omega$ and that the field is linearly polarized, then second-order perturbation theory yields

$$w_{pn} = 2\left|\chi_{np}^{(1)} E_\omega^2\right|^2 \Gamma_p \left[\left(\omega_{pn} - 2\omega\right)^2 + \Gamma_p^2\right]^{-1}, \qquad (2.31)$$

where the two-photon matrix element of the $n \to p$ transition is given by a formula corresponding to the expression (1.21) for the off-diagonal matrix element for the linear susceptibility:

$$\chi_{np}^{(1)} = \sum_m z_{nm} z_{mp} \left[\left(\omega_{mn} - \omega\right)^{-1} + \left(\omega_{mn} + \omega\right)^{-1}\right]. \qquad (2.32)$$

If a one-photon resonance is present ($\omega \approx \omega_{mn}$), the polarization of the medium is determined by formulas similar to (2.28), but instead of the two-photon matrix element we have the simpler one-photon matrix element z_{mn} (for detailed calculations refer to Rautian, Smirnov, and Shalagin, 1979, Section 7). Since

$$N_n - N_p = \left(1 + \frac{w_{pn}}{\Gamma_p}\right)^{-1}, \qquad (2.33)$$

for $w_{pn} \gg \Gamma_p$ (that is, in a strong field) the difference in the populations vanishes, which leads to saturation of the $n \to p$ transition by a strong light field.

2.2.4. Conclusion

Thus, the resonance nonlinear susceptibility determines, just as the nonresonance nonlinear susceptibility does, the nonlinear scattering of light by an
atom. It gives the cross sections of the corresponding resonance processes:
either the elastic resonance nonlinear scattering (when the final state of the
atom coincides with the initial) or the inelastic resonance nonlinear scattering (when the final state differs from the initial). For a comparison of these
processes with resonance linear scattering, we can repeat here all the
conclusions made at the end of Section 2.1.

A resonance in the nonlinear susceptibility leads to a factor $z_{mn}E_\omega/\Gamma_m$,
where Γ_m is the width of state m (see Section 1.2), in the induced dipole
moment of the atom. If the light field is strong, these factors may exceed
unity considerably. In this way they can to some extent compensate the
smallness of the nonresonance factors E_ω/E_{at} in the dipole moment of the
atom, and the atomic polarization may reach large values, which can be
observed in experiments. Thus, one must always try to realize the maximum
possible number of resonances and establish small widths of the resonance
states.

2.3. HYPERPOLARIZABILITY OF ATOMIC LEVELS

In Section 1.1.4 we discussed the dynamic polarizability of atoms that leads
to Stark shifts in the atomic levels, shifts quadratic in the electric field
strength E of the external light wave. We found that the proportionality
factor in the dependence of the Stark shift $\delta\mathscr{E}_n$ on the intensity $I_\omega \propto E_\omega^2$ is
the linear atomic susceptibility $\chi_{nn}^{(1)}(\omega; \omega)$ (or polarizability). Here ω is the
frequency of the light wave.

In the interresonance regions, as we have seen, $\chi^{(1)}$ may vanish. Then
one must take into account the next terms in the expansion of $\delta\mathscr{E}_n$ in
powers of the intensity I_ω of the light wave, that is, terms proportional to
I_ω^2. The proportionality factor will then be determined by the nonlinear
(cubic) susceptibility $\chi_{nn}^{(3)}(\omega; \omega, -\omega, \omega)$, which is commonly known as the
hyperpolarizability of the atom in the given state n when a light field of
frequency ω acts on the atom.

In contrast to the linear case, where the polarizability of the atom, as we
saw earlier, reduces completely to the linear susceptibility $\chi_{nn}^{(1)}(\omega; \omega)$, in the
nonlinear case the hyperpolarizability of the atom contains another term in
addition to $\chi_{nn}^{(3)}$. If we consider the expression (1.13) for the quadratic Stark
shift in a variable field (here we assume for simplicity that E_ω is directed

along the z axis and is linearly polarized), we have

$$\delta \mathscr{E}_n^{(2)} = \sum_{m \neq n} |z_{nm}|^2 E_\omega^2 \left[(\omega_{mn} - \omega)^{-1} + (\omega_{mn} + \omega)^{-1} \right]. \quad (2.34)$$

The term in the Stark shift proportional to E_ω^4 appears if we allow in this formula for the corrections to the denominators $\omega_{mn} - \omega$ and $\omega_{mn} + \omega$, corrections that are simply differences between the Stark shifts of levels m and n, that is, $\omega_{mn} \pm \omega$ are replaced by $\omega_{mn} + \delta \mathscr{E}_m^{(2)} - \delta \mathscr{E}_n^{(2)} \pm \omega$. This procedure corresponds to the perturbation-theory expansions with exact values for the energy in the propagators. After expanding these propagators in powers of $\delta \mathscr{E}_m^{(2)}$ and $\delta \mathscr{E}_n^{(2)}$, we obtain

$$\left[\omega_{mn} \pm \omega + \delta \mathscr{E}_m^{(2)} - \delta \mathscr{E}_n^{(2)} \right]^{-1}$$
$$\approx (\omega_{mn} \pm \omega)^{-1} - (\omega_{mn} \pm \omega)^{-2} \left(\delta \mathscr{E}_m^{(2)} - \delta \mathscr{E}_n^{(2)} \right). \quad (2.35)$$

As a result, if we substitute this expression into (2.34), we find the contribution to $\delta \mathscr{E}_n^{(4)}$:

$$E_\omega^2 \sum_{m \neq n} |z_{mn}|^2 \left[\delta \mathscr{E}_n^{(2)} - \delta \mathscr{E}_m^{(2)} \right] \left[(\omega_{mn} + \omega)^{-2} + (\omega_{mn} - \omega)^{-2} \right]. \quad (2.36)$$

Since $\delta \mathscr{E}_m^{(2)}$ and $\delta \mathscr{E}_n^{(2)}$ are proportional to E_ω^2, this contribution is proportional to E_ω^4 and constitutes the sought additional term in the Stark shift $\delta \mathscr{E}_n^{(4)}$, but cannot be reduced to the cubic nonlinear susceptibility $\chi^{(3)}(\omega; \omega, -\omega, \omega)$. The final expression for the hyperpolarizability is not given here, as it is too cumbersome (see Delone and Kraĭnov, 1985, Chapter 6, and Rapoport, Zon, and Manakov, 1978).

At first glance it would seem that the higher-order terms in the expansion of the atomic level shift in powers of the radiation intensity I_ω become comparable with the first term in the expansion when the field strength E_ω is of the order of the atomic field strength E_{at}. But calculations have shown this not to be so. For instance, in alkali atoms these terms are of the same order of magnitude at $E \sim 10^{-3} E_{at}$ (see Rapoport, Zon, and Manakov, 1978, Section 3.3). There are two reasons for this. First, the denominators in the nonlinear susceptibility are not of the order of unity but are severalfold smaller; besides, the susceptibility contains a product of several denominators. Second, the dipole matrix elements between the excited virtual states of an atom, the elements present in the numerator of the nonlinear susceptibility, are usually severalfold greater than unity. As we have seen, the susceptibility contains the product of several dipole matrix elements.

2.4. MULTIPHOTON EXCITATION AND NONLINEAR IONIZATION OF ATOMS

2.4.1. Introduction

In discussing the role of the imaginary part of the resonance nonlinear susceptibility we saw that ionization broadening constitutes one mechanism of broadening (see Section 2.2.3). This effect is caused by both single-photon and multiphoton ionizations, depending on the relationship between the frequency of the radiation field and the energy of the resonance level. For instance, in the case of third-harmonic excitation, after the atomic electron has absorbed three photons from the radiation field of frequency, it finds itself in a highly excited state, where it can absorb another photon of frequency ω (or several such photons) and go over to the continuous spectrum instead of emitting a photon of frequency 3ω and returning to the initial state n. We will discuss the aspects that make ionization competitive with third-harmonic excitation or other nonlinear optics processes in Chapter 3. Here we only discuss the process of nonlinear ionization of an isolated atom by a strong light field. Another mechanism of broadening connected with the resonance nonlinear susceptibility is caused by spontaneous transitions into the initial state (by one or several transitions) after the atom has been excited resonantly into a state in the discrete spectrum via absorption of several photons from the external field.

Earlier in this chapter we dealt primarily with nonlinear susceptibilities related to the diagonal matrix elements of the atomic dipole moment. Off-diagonal matrix elements were encountered only in the description of resonance nonlinear susceptibility [see Eq. (2.30)]. In this section, however, we are dealing with off-diagonal multiphoton matrix elements, since physically the initial state in the ionization process belongs to the discrete spectrum, while the final state belongs either to the continuous spectrum or to another state in the discrete spectrum in the event of multiphoton excitation—that is, the two states differ. In all other respects the rules for constructing Feynman diagrams responsible for the abovementioned multiphoton matrix elements remain the same as those formulated at the beginning of this chapter, that is, they are based on the higher-order quantum perturbation theory.

2.4.2. Conditions Necessary for Multiphoton Ionization

Before we discuss the process of multiphoton ionization using higher-order quantum perturbation theory, let us see under what conditions such a process occurs.

As we have already said in Chapter 1, the photoionization of atoms, that is, one-photon ionization, cannot occur for atoms in the ground state or for fields in the optical range, since then $\mathscr{E}_n$ is always greater than $\hbar\omega$. The situation is just the opposite when there is nonlinear ionization. Such a process can be realized for any energy of an electron in the atom, $\mathscr{E}_n$, and any frequency of the radiation field, $\hbar\omega < \mathscr{E}_n$, since the energy conservation law is always satisfied by the atom absorbing a sufficient number of photons.

The general theory of nonlinear ionization of atoms in a monochromatic electromagnetic field, a theory built on the model of electron emission caused by a delta-like potential (see Keldysh, 1964, and also Delone and Kraĭnov, 1985, Chapter 6) with a single bound state, shows that the ionization process is determined by a dimensionless quantity known as the *adiabaticity parameter* ($e = M_e = 1$):

$$\gamma = \frac{\omega(2\mathscr{E}_n)^{1/2}}{E_\omega}. \tag{2.37}$$

For $\gamma \ll 1$ ionization occurs due to the tunneling of the electron through a quasistationary barrier, while for $\gamma \gg 1$ this process occurs due to absorption of $K = \lfloor \mathscr{E}_n/\hbar\omega + 1 \rfloor$ photons by the electron, which is necessary for energy conservation when the electron goes over to the continuous spectrum (Kth-order perturbation theory). Here the floor brackets stand for the integral part of the number inside of them. Although the real potential of an optical electron in an atom differs drastically from a delta-like potential and the spectrum of bound states contains an infinite number of levels instead of one, the entire body of experimental data on nonlinear ionization of atoms known to us confirms these general conclusions derived from the theory (Delone and Kraĭnov, 1985).

The formula (2.37) for the adiabaticity parameter shows that in the case of interest to us, of relatively high radiation frequencies of the optical range, where the frequency ω of the radiation field is not too low compared with the frequency $\mathscr{E}_n/\hbar$ of the transition to the continuous spectrum and the radiation field strengths are not too high ($E_\omega \ll E_{\mathrm{at}}$), ionization is a multiphoton process. For this reason we will now discuss only the multiphoton ionization of atoms and will not consider either the tunneling effect in a variable field or the intermediate case $\gamma \sim 1$.

2.4.3. Multiphoton Ionization of Atoms

The nature of the process of multiphoton ionization of atoms depends largely on whether there are intermediate resonances between the energy of the bound state of an electron in the atom and the energy of several photons from the radiation field. If resonances are absent, the multiphoton ionization process is called *direct*, while if resonances are present, the process is called *resonance multiphoton ionization*.

The conditions for the realization of direct multiphoton ionization are that all misfits of the intermediate resonances must be much larger than the widths of the corresponding resonance states, Γ_m, which can be written as

$$\Delta = |\omega_{mn} + \delta\mathscr{E}_m - \delta\mathscr{E}_n - K'\omega| \gg \Gamma_m. \qquad (2.38)$$

Here n is the ground state, m the intermediate resonance state in the spectrum of the atom, and K' the number of photons that make possible the particular resonance. Obviously, Γ_m takes into account the perturbation of the atomic levels in the radiation field (see the mechanisms of broadening of resonance levels discussed in Section 1.2.4). In a weak radiation field there is no need to take into account either the Stark shifts of levels n and m in a variable field or the effect of the radiation field on the width. The choice of state m in (2.38) is determined by the selection rules for multiphoton transitions (Delone and Kraĭnov, 1985), rules based on the dipole selection rules for one-photon transitions. Note that in a sufficiently strong field these rules may change (see Delone and Kraĭnov, 1985, Section 1.3).

When the condition (2.38) is met, that is, when direct multiphoton ionization is realized, the probability of this process (its rate, to be more precise) is determined by the Fermi "golden rule," or the following power-function relationship:

$$w_n = |\chi_{np}^{(K)}E_\omega^K|^2 p(2\pi)^{-2}, \qquad (2.39)$$

where p is the electron's momentum in the final state, and $\chi_{np}^{(K)}$ is the off-diagonal multiphoton matrix element in the Kth-order perturbation theory. Figure 2.16 shows the process of absorption of K photons from the radiation field, while Figure 2.17 gives the Feynman diagram for $\chi_{np}^{(K)}$. The energy conservation rule then has the form $K\omega = \mathscr{E}_p - \mathscr{E}_n$. Note that the final state lies in the continuous spectrum. The coefficient of E_ω^{2K} on the right-hand side of (2.39) is known as the multiphoton cross section and is denoted by $\alpha^{(K)}$. This quantity, obviously, depends on the frequency of the radiation field and the polarization of this field.

The condition for realization of the resonance ionization process is simply opposite to (2.38), namely,

$$\Delta < \Gamma_m. \qquad (2.40)$$

In a weak light field, whose effect on the width of resonance level m one can ignore, the probability of resonance ionization is obtained from (2.39)

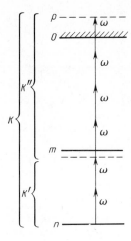

Figure 2.16. Absorption of photons in multiphoton resonance ionization of an atom; m is the resonance intermediate level, K is the total number of absorbed photons, and K' is the number of photons absorbed during resonance excitation of level m.

by applying the Breit–Wigner procedure to the resonance propagator, or $\Delta \rightarrow \Delta - i\Gamma_m$. We then have

$$w_n = N_m w_m. \qquad (2.41)$$

Here N_m is the absolute probability of population of resonance state m:

$$N_m = \left|\chi_{nm}^{(K')} E_\omega^{K'}\right|^2 \left[\left(\omega_{mn} - K'\omega\right)^2 + \Gamma_m^2\right]^{-1} \qquad (2.42)$$

(the population of state m due to the radiation field). Here $\chi_{nm}^{(K')}$ is the matrix element of the multiphoton transition from ground state n to

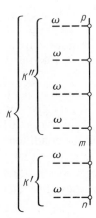

Figure 2.17. The Feynman diagram for the off-diagonal matrix element of the nonlinear susceptibility corresponding to the process depicted in Figure 2.16.

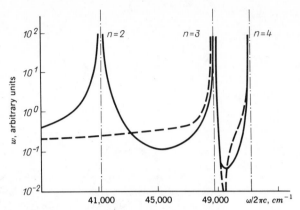

Figure 2.18. The probability of three-photon ionization per unit time (arbitrary units) of the ground state of a hydrogen atom as a function of the frequency of the electromagnetic wave. The solid curves correspond to a linearly polarized wave, and the dashed curves to a circularly polarized wave. At $n = 2$ there is no resonance for the circularly polarized wave, thanks to the selection rule in the magnetic quantum number. All resonances are two-photon.

resonance state m (determined in the same manner as $\chi^{(K)}$), and K' is the number of photons necessary for the resonance transition from n to m (see Figure 2.17).

The quantity w_m in (2.41) is the probability of K''-photon ionization ($K' + K'' = K$) of the resonance level m per unit time (the rate of this ionization); it is determined by a formula similar to (2.39):

$$w_m = \left| \chi_{np}^{(K'')} E_\omega^{K''} \right|^2 p(2\pi)^{-2}. \tag{2.43}$$

For intense light (that is, when we must allow for the perturbation of the resonance state by the light field) and for nonmonochromatic radiation, the formulas for the resonance multiphoton ionization probabilities have a more complex form (see Basov, 1980, and Delone and Kraĭnov, 1985) and will not be given here.

To illustrate the general behavior of the multiphoton ionization of an atom with respect to the frequency of the light, in Figure 2.18 we give the results obtained by Zon, Manakov, and Rapoport, 1972, for the case of three-photon ionization of a hydrogen atom. Figure 2.18 and the conditions for the realization of direct ionization (2.38) and resonance ionization (2.40) show that the direct process is usually realized, while the resonance process requires special measures for selecting the radiation frequency. For this reason we will only discuss direct multiphoton ionization.

2.4.4. Direct Multiphoton Ionization of Atoms

The cross sections of direct multiphoton ionization, as functions of the frequency and polarization of the radiation, have been studied in detail both experimentally and by applying time-dependent perturbation-theory methods for many atoms (Basov, 1980; Delone and Kraĭnov, 1985; and Rapoport, Zon, and Manakov, 1978). To generalize the results of these studies and describe direct multiphoton ionization, we specify three main characteristics of the process: the absolute value of multiphoton cross sections, their frequency dependence, and their dependence on the polarization of the light.

The absolute values of multiphoton ionization cross sections, measured in experiments and calculated theoretically, are in satisfactory agreement for all atoms studied, that is, alkali atoms (with one optical electron), alkaline-earth atoms (two optical electrons), and a number of other atoms including those of noble elements (completely filled outer shell). Data have been gathered for degrees of nonlinearity (the number K of absorbed photons) ranging from 2 to 11. The most thoroughly studied are the alkali and alkaline-earth atoms for low degrees of nonlinearity ($K \leq 5$), and this case is the most interesting for nonlinear optics. Note, moreover, that the statement that the experimental and theoretical data agree fairly well takes into account the fairly high inaccuracy in measurements of multiphoton ionization cross sections, which is due to the methods of measuring these cross sections (Delone and Kraĭnov, 1985; Basov, 1980). For instance, the mean error in the cross section is of the order of the quantity itself (at medium degrees of nonlinearity). In Table 2.2 we list the averaged cross sections for different degrees of nonlinearity K. Note that (2.39) implies that the dimensions of multiphoton cross sections depend on the degree of nonlinearity, $[\alpha^{(K)}] = \mathrm{cm}^{2K}\mathrm{s}^{K-1}$, which means there is no point in comparing cross sections with different K's. (The values of the cross sections in Table 2 are averaged over the numerous results of experiments and calculations, and the accuracy is about one order of magnitude. The values are

Table 2.2   Cross Sections $\alpha^{(K)}$ of Direct Multiphoton Ionization of Atoms

Unit	$-\log_{10} \alpha^{(K)}$					
	$K = 2$	3	4	5	6	7
$\mathrm{cm}^{2K}\,\mathrm{s}^{(K-1)}$	48	78	106	139	170	204
Atomic units	-2	-4	-9	-10	-12	-12

actually $-\log_{10} \alpha^{(K)}$, and the atomic unit of $\alpha^{(K)}$ is $a^{2K}\tau^{K-1}$, where a is the Bohr radius and $\tau = \hbar^3/me^4$ is the atomic time.) What should be compared are the ionization probabilities (2.39) at a fixed field strength E_ω.

The observed dependences of the multiphoton cross sections of direct ionization on the light's frequency are the simplest in the case of ionization of alkali atoms. Resonances are unambiguously interpreted as those between the energy of the transition of an electron from the ground state to a definite excited state and the energy of several photons from the radiation field. In a strong field there is a dependence of the resonance frequency on the field strength, which is caused by the Stark shift of atomic levels due to the radiation field (Delone and Kraĭnov, 1985; Zon, Manakov, and Rapoport, 1972). Beterov, Fateev, and Chebotaev, 1982, also found that there are resonances in the ionization probability caused by nonresonance excitation of bound electronic states, but this case is very unusual. Resonances were also observed in the ionization of alkaline-earth atoms. Such resonances are caused by two-electron bound states and autoionization states (Delone, 1985).

The dependence of the multiphoton cross sections on the polarization of the light for atoms with one valence electron and $K \leq 5$ has certain characteristic features; in particular, the ionization probability for circularly polarized light, w_C, is always higher than the same quantity for linearly polarized light, w_L. For the ratio of these probabilities there is the following formula (Basov, 1980; Delone and Kraĭnov, 1985; and Klarsfeld and Maquet, 1972):

$$\frac{w_C}{w_L} = \frac{(2K-1)!!}{K!}. \tag{2.44}$$

The behavior just discussed is caused by the fact that the probability of the virtual transitions of the electron over the bound-state spectrum in the course of photon absorption is highest when an increase in the principal quantum number by unity is accompanied by the same increase in the orbital quantum number. This is a well-known behavior for transitions in the spectrum of the hydrogen atom (Bethe and Salpeter, 1957). For alkali atoms, Eq. (2.44) has been substantiated by many experiments (Basov, 1980; Morellec, Normand, and Petite, 1982). In the case of ionization of atoms with several valence electrons, Eq. (2.44) does not hold true, as is shown by the example of alkaline-earth atoms. This results from the fact that the spectrum of atoms with many electrons in the outer shell is not hydrogenlike.

In conclusion we note that in the event of multiphoton ionization of atoms with several electrons in the outer shell, many-electron ionization is also observed (Delone and Kraĭnov, 1985), in addition to the main process of one-photon ionization. However, the probability of this process occurring is much lower than that for one-electron ionization, so that many-electron ionization plays no role in nonlinear-optics phenomena.

2.4.5. Probability of Multiphoton Excitation

If during the absorption of several photons there occurs resonance multiphoton excitation of a certain atomic state, this excited state may actually get populated, provided that there is exact tuning to resonance, and instead of stimulated absorption of the next group of photons the state may be deexcited by emission of one or several spontaneous photons emitted in the form of cascades. This process leads to nonlinear absorption of the light in a gaseous medium. Chapter 3 is devoted to this. Here we give only the expression for the probability of multiphoton excitation of an atom from state n into state m, which in accord with the Fermi "golden rule" has the form

$$w_{nm} = 2\pi \left| V_{mn}^{(K)} \right|^2 \rho_m, \qquad (2.45)$$

where ρ_m is the Lorentzian function that determines the spontaneous decay of the excited resonance state m, and $V_{mn}^{(K)}$ is the multiphoton matrix element that determines the $n \rightarrow m$ transition accompanied by absorption of K photons from the external light field. The law of energy conservation must also be satisfied, that is, $\omega_{mn} = K\omega$, and this ensures that the state m is resonantly populated in the course of multiphoton excitation.

Moreover, for the multiphoton matrix element $V_{mn}^{(K)}$ to be nonzero, certain selection rules must be satisfied: the parity of state m must coincide with that of state n if K is even, while the parities of m and n must be opposite if K is odd. Since the absorption of a single photon changes the orbital angular momentum l of the state by unity (in the one-electron approximation, which is assumed to be valid throughout), for even K we have $l_m = l_n, l_n \pm 2, \ldots, l_n \pm K$, while for K odd we obtain, similarly, $l_m = l_n \pm 1, l_n \pm 3, \ldots, l_n \pm K$. As for the selection rules in magnetic quantum numbers, they depend on the polarization of the external light field and are treated in detail in Delone and Kraĭnov, 1985, Section 1.3, where the selection rules in the spin and total angular momentum are also given (these differ for the cases of weak and strong light field and depend on the type of coupling of electron spin with the orbital angular momentum of the electron in the atom).

The multiphoton matrix element $V_{mn}^{(K)}$ has the following form (for simplicity we assume that the external light field with an electric field strength $2E_\omega \cos \omega t$ is linearly polarized):

$$V_{mn}^{(K)} = E_\omega^K \sum_{s,p,f,\ldots,q} z_{ns} z_{sp} z_{pf} \cdots z_{qm}$$

$$\times \left[(\omega_{sn} - \omega)(\omega_{pn} - 2\omega) \cdots (\omega_{qn} - (K-1)\omega) \right]^{-1}. \quad (2.46)$$

This can also be written in terms of the nonlinear resonance polarizability in the following form:

$$V_{mn}^{(K)} = \chi_{mn}^{(K-1)}((K-1)\omega; \omega, \omega, \ldots, \omega)), \qquad (2.47)$$

which is similar to the expression for the case of multiphoton ionization discussed above.

2.5. CONCLUSION

If we compare the main features of nonlinear and linear interactions of light with an atom, the main conclusion to be drawn is that there is a much broader spectrum, so to say, of processes connected with nonlinear interaction than with linear interaction. First, photons of quite different frequencies can be formed during a nonlinear interaction, frequencies that may exceed the frequency of the exciting light many times over. This explains the nonlinear process of excitation of optical harmonics, which is basic to nonlinear optics. It is considered in Section 3.2. Second, the average optical characteristics of the medium change. This is true, in particular, of the index of refraction of the medium, which explains the nonlinear refraction of the medium. This phenomenon is considered in Section 3.5. Finally, since nonlinear polarization manifests itself at frequencies that differ from the frequency of the exciting light, the law of additivity of light fluxes breaks down and waves of different frequencies can couple through the nonlinear polarization (see Section 3.4). The basic macroscopic consequences of the nonlinear interaction of strong light with an atom are considered in Chapter 3.

We note, first, the changes that occur when we go over from atoms to molecules. In contrast to atoms, there are vibrational and rotational states in the spectra of bound states of molecules. The separation between neighboring vibrational terms is much smaller than that between electron levels (the energies of electron levels in molecules are of the same order of magnitude as in atoms). Further, the separation between neighboring rotational terms is considerably smaller than that between neighboring vibrational terms. Thus, the energies of vibrational transitions, and even more the energies of rotational transitions (and mixed rovibronic transitions), prove to be very small compared with the energy of a photon from the optical frequency range. As a result there can be no resonance between the frequency of visible light and the frequencies of a vibrational or rotational transition.

In the second place, in determining the nonlinear susceptibility of a molecule we must ignore the vibrational and rotational terms, just as in determining the nonlinear atomic susceptibility we ignored the hyperfine structure of the atomic levels. This is true in particular when we are studying processes with transitions to all the vibrational and rotational terms belonging to a given electronic state, that is, summation has been carried out over the intermediate vibrational and rotational states. If we wish, say, to know the probability of transition from a given vibrational or rotational term, this can be found if we know the probability of transition from the corresponding electronic state, which must then be multiplied by a geometric factor (a detailed study of this procedure can be found in Berestetskiĭ, Lifshitz, and Pitaevskiĭ, 1980, §61).

The next difference between a molecular gas and an atomic gas lies in the fact that molecules can possess a constant dipole moment, and hence the nonlinear susceptibility $\chi^{(2)}$ is nonzero for them (Section 2.1.1).

For the nonlinear interaction of light with atomic and molecular ions, the one and only special feature is that, in contrast to neutral particles, charged particles may vibrate as a whole in the field of a light wave (in addition to changes in the internal state). These vibrations are classical rather than quantum. The same vibrations are experienced in the field of a light wave by free electrons formed in the process of ionization in a gaseous medium.

3

Nonlinear-Optics Phenomena in Atomic Gases

In this chapter we will consider the main nonlinear effects appearing in the interaction of intense light with an atomic gas, namely, the changes in frequency, polarization, and direction of propagation, the nonlinear refraction and absorption of the incident light, and the phenomenon of nonlinear reflection from the interface between the atomic gas and vacuum. These effects constitute the basis of nonlinear optics in other media, too.

At present nonlinear optics is a well-studied and deeply developed section of physics. For this reason, in the scientific periodical literature, in numerous reviews, and in monographs cited in the Introduction to this book, one may find detailed descriptions of the nonlinear-optics phenomena in various media and under different assumptions concerning the properties of the incident radiation, that is, the temporal and spatial distribution of the radiation intensity, the curvature of the incident wavefront, and the degree of monochromaticity of the radiation.

As has been said in the Introduction, our goal is not to describe the entire scope of nonlinear-optics phenomena. It is more modest—we wish to discuss the main phenomena in the simplest possible case, starting from the microscopic effects on the atomic level and going to macroscopic effects. For this reason the discussion in this chapter (like the one concerning linear optics in Chapter 1) is carried out within the scope of various approximations and simplifying assumptions.

In the Introduction we substantiated the models, approximations, and simplifying assumptions within whose framework we discussed the linear and nonlinear effects. Since lasers are the real source of intense light, the simplifying assumptions and approximations are discussed and substantiated below from the standpoint of the real properties of laser radiation.

3.1. THE BASICS OF THE THEORY OF THE INTERACTION OF INTENSE LIGHT WITH AN ATOMIC GAS

3.1.1. Introduction

Since in this chapter we speak of the interaction of light waves with a macroscopic medium (wave optics), the main conclusions will be obtained by solving Maxwell's equations and illustrated by employing the language of the geometrical optics of light rays. Here we are interested in nonlinear effects, and so the main characteristic of a gaseous medium will be its nonlinear susceptibility, the quantity that determines its nonlinear polarization in the field of a light wave. This explains the nonlinearity of Maxwell's equations for this case. The time-independent solutions of the nonlinear Maxwell equations will be obtained under certain boundary conditions, connected both with the general formulation of the problems considered and the adopted simplifying assumptions and approximations. We start our discussion with these assumptions and approximations.

3.1.2. Simplifying Assumptions and Approximations

The simplifying assumptions and approximations formulated below have already been briefly discussed in the Introduction to this book and at the beginning of Chapter 1. However, many of these are not necessary in discussing problems of linear optics. In this chapter these approximations are used in considering the nonlinear process of the interaction of light waves with a macroscopic medium and for this reason are discussed in greater detail. In the rare cases where we go outside the scope of these approximations, we will specify this. We start with the characteristics of light waves.

3.1.2.1. The Monochromaticity of Light Waves. We will assume that light waves are ideally monochromatic, that is, $\Delta\omega/\omega = 0$, where $\Delta\omega$ is the width of the radiation spectrum. This assumption is quite realistic, since we are speaking of laser radiation, for which $\Delta\omega/\omega$ varies between 10^{-3} and 10^{-8}. The real quasimonochromaticity of laser radiation can play an essential role only when the interaction of light with an atom has a resonance nature. The data on nonlinear susceptibilities $\chi^{(K)}$ from Chapter 2 show that although $\chi^{(K)}$ is always a function of frequency ω, far from resonances this function is constant on the scale of $\Delta\omega$. In the presence of resonances, however, the real quasimonochromaticity of laser radiation may play an essential role. For this reason, in some particular cases the results obtained on the assumption of the monochromaticity of the radiation may give an incorrect description of the real situation. We will allow for the quasimonochromaticity of laser radiation when this is necessary for comparing theoretical predictions with experimental data.

3.1.2.2. The Plane Incident Wavefront.

We will assume that the wavefront of the wave incident on the nonlinear medium is plane. This assumption is quite realistic, since the laser-beam divergence is usually determined by diffraction, and therefore it is of the order $\lambda/a \sim 10^{-4}$, with a the radius of the laser beam. In some cases, however, we must allow for the divergence of the laser beam, for instance, in describing the restrictions on the self-focusing of laser radiation.

3.1.2.3. Uniform Distribution of Field Strength over the Incident Wavefront.

This assumption is employed to simplify the mathematics. It is also assumed that the wave is unlimited in the plane of the wavefront. Thus, the three-dimensional problem (or two-dimensional in the case of axial symmetry in the distribution) is reduced to a one-dimensional one. This assumption, however, is unrealistic. Indeed, the distribution of the field strength over the wavefront of a laser beam is usually determined by diffraction and is therefore essentially nonuniform. A good approximation here is a Gaussian distribution over the transverse coordinate r of the cylindrical system of coordinates r, z, ψ. The simplifying assumption of the uniform distribution can be employed only because in the majority of cases the distribution over r plays almost no role. When this is not so, for instance, in describing nonlinear refraction, we will allow for the real nonuniform distribution.

3.1.2.4. The Stationary Action of the Light Radiation.

As in the previous chapters, all reasoning in this chapter is conducted on the assumption that the interaction of light with a medium is time-independent, or stationary. At first glance this is completely unrealistic, since the large intensities of the light fluxes we are interested in are obtained in pulsed lasing, with a typical pulse length T of the order 10^{-10} to 10^{-8} s (mode locking and Q switching). However, in most cases the radiation, even in the form of such short pulses, is equivalent to a stationary interaction. These are the cases where the response time τ of the medium is much shorter than the pulse length. The condition $\tau \ll T$ obviously is satisfied for all nonresonance processes. But when there are resonances and the atomic levels may become really occupied, $\tau \sim \Delta^{-1}$ may become much larger than T (where Δ is the deviation from resonance) and the interaction cannot be considered to be stationary. A great variety of nonstationary nonlinear-optics effects are known that differ substantially from stationary effects. As an example we can point to the effect of self-induced transparency.

3.1.2.5. Polarization of the Radiation. We will assume that the radiation is completely polarized. This assumption is quite realistic, since in the majority of lasers the radiation is completely polarized, and when this is not so, radiation with a given polarization can always be separated from a beam of laser radiation.

3.1.2.6. The Unperturbed-Field Approximation. A brief description of this approximation was given in the Introduction to this book. Let us consider the specific features of applying this approximation in nonlinear optics. Obviously, if we allow for the higher-order terms in the expansion of the polarization in powers of the field strength, Maxwell's equations are generally nonlinear (Akhmanov and Khokhlov, 1964; Bloembergen, 1965). An analytical solution of these equations is possible only in the simplest cases (Whitham, 1974). However, most physical processes can be understood and described using this approximation when the excited electromagnetic waves, known as polarization waves, are weak compared to the external electromagnetic waves incident on the medium and creating nonlinear polarization in the medium. This explains why the changes occurring in the incident wave can be neglected in determining the polarization. The unperturbed-field approximation corresponds to the condition $E_\nu \ll E_\omega$, where $\mathbf{E}_\nu$ and $\mathbf{E}_\omega$ are the maximum field strengths of the excited and incident waves, respectively, and ν and ω are the frequencies of these waves.

Thus, we have arrived at a much simpler problem, in which Maxwell's equations for the weak excited electromagnetic wave are considered in the approximation of a given (unperturbed) incident electromagnetic wave (or several such waves), which results in nonlinear polarization. Such an approximation is customarily known (as noted at the beginning of this book) as the unperturbed-field approximation. The corresponding Maxwell equations are linear inhomogeneous partial differential equations, and the inhomogeneous term is the nonlinear polarization of the medium taken as a given function of coordinates and time. In the subsequent sections, when discussing concrete physical processes, we will consider the restrictions of physical parameters under which the unperturbed-field approximation is realized.

Now let us turn to the properties of the atomic medium.

3.1.2.7. The Absence of an Interaction between the Atoms. It is assumed that atoms and other microscopic particles (electrons and ions) constituting the medium and forming in it as a result of radiation do not interact with each other in the course of T, the pulse length; that is, there are no collisions in the course of T. This is an important condition, since only if it is met can we assume the unperturbed atomic spectrum in the case of a weak external field, or the atomic spectrum perturbed only by the external field (that is, the spectrum of the quantum system consisting of the atom and the field—the *dressed atom*) in the case of a strong external field. If during the time that the external field is acting there are collisions between the atoms (known as radiative collisions; see Yakovlenko, 1984), then the spectrum of the particles participating in the collisions in the field differs from the collisionless spectrum.

The condition of absence of interaction between the atoms during the laser pulse imposes an obvious upper bound on the density of the atomic gas. The numerical value of this bound depends on the composition and other properties of the atomic medium. The limiting value of the concentration is always less (in order of magnitude) than 10^{16} cm^{-3}.

3.1.2.8. The Immobility of the Atoms.

We do not consider the motion of the atoms, thus assuming them to be immobile. In a real situation the atoms are in thermal motion, which leads to Doppler absorption-line broadening (see Section 1.2.4). Doppler broadening, like other broadening mechanisms, plays no role in nonresonance processes and must be taken into account only when resonances appear.

The above-noted simplifying assumptions and approximations form the basis for studying the process of nonlinear interaction of intense light with an atomic medium. This is done in the subsequent sections. In all cases when for various reasons we step outside of these simplifying assumptions and approximations, we will specify this.

3.1.3. The Maxwell Equations in a Nonlinear Medium

Note that in this section we are not interested in how the incident electromagnetic waves which result in nonlinear polarization propagate in the medium. Their propagation is described by linear Maxwell equations of macroscopic electrodynamics in accordance with the appropriate boundary conditions at the interface of the gaseous medium considered (the waves propagate in this medium). Our aim is only to describe the creation of new electromagnetic waves, that is, waves whose frequencies differ from those of the incident external waves, or to describe the nonlinear changes in the incident external waves at the same frequencies.

In accordance with the results obtained in Chapter 2, the nonlinear part of the atomic polarization created by one or more incident monochromatic electromagnetic waves with field strengths $\mathbf{E}_i \exp(i\omega_i t - i\mathbf{k}_i \cdot \mathbf{r})$, with subscript i numbering the incident waves, can be written in the following general form:

$$\mathbf{P}^{(K)} = \mathbf{P}_\nu^{(K)} \exp(i\nu t - i\mathbf{k} \cdot \mathbf{r}). \qquad (3.1)$$

The amplitudes $\mathbf{P}^{(K)}$ have the following form in the unperturbed-field approximation:

$$\mathbf{P}_\nu^{(K)} = \chi^{(K)}(\nu; \omega_1, \omega_2, \ldots, \omega_K)\mathbf{E}_1\mathbf{E}_2 \cdots \mathbf{E}_K, \qquad (3.2)$$

where $\chi^{(K)}$ is the nonlinear atomic susceptibility (per atom), and we have not taken into account the tensor structure of the nonlinear susceptibility. The frequency ν and the wave vector $\mathbf{k}$ are given by the following formulas:

$$\nu = \sum_{i=1}^{K} \omega_i, \qquad \mathbf{k} = \sum_{i=1}^{K} \mathbf{k}_i, \qquad (3.3)$$

and are the sums of the frequencies and wave vectors of the waves incident on the nonlinear medium.

Let us denote the electric field strength in the nonlinear polarization wave by $\mathbf{E}'(\mathbf{r}, t)$. In accordance with what was said in Section 3.1.2, the Maxwell equation for the polarization wave at frequency ν can be written (see Born and Wolf, 1975) as follows:

$$\nabla^2 \mathbf{E}'(\mathbf{r}, t) - \left(\frac{n_\nu}{c}\right)^2 \frac{\partial^2 \mathbf{E}'(\mathbf{r}, t)}{\partial t^2} = 4\pi \frac{N}{c^2} \frac{\partial^2 \mathbf{P}^{(K)}}{\partial t^2}, \qquad (3.4)$$

where ∇^2 is Laplace's differential operator, $\mathbf{P}^{(K)}$ is defined in (3.1) and (3.2), and N is the number of atoms per unit volume of the gas. The above equation determines the nonlinear part of the polarization, while the linear part is taken into account by introducing the appropriate refractive index n_ν at frequency ν. This index is close to unity in gaseous media. The fact that Eq. (3.4) is written in the unperturbed-field approximation means that E' is much lower than E_i.

Equation (3.4) is a linear inhomogeneous differential equation for $\mathbf{E}'(\mathbf{r}, t)$. The right-hand side of this equation is the given inhomogeneous term.

Combining (3.1) and (3.4), we can isolate the dependence of the field strength in the polarization wave on time:

$$\mathbf{E}'(\mathbf{r}, t) = \mathbf{E}'(\mathbf{r})\exp(i\nu t). \qquad (3.5)$$

Thus, a weak electromagnetic polarization wave of frequency ν is excited, and the nonlinear polarization vector $\mathbf{P}^{(K)}$ oscillates with the same frequency. The vector $\mathbf{E}'(\mathbf{r})$ is the amplitude of the polarization wave.

Substituting (3.1) and (3.5) into (3.4), we arrive at an equation for the amplitude of the polarization wave:

$$\nabla^2\mathbf{E}'(\mathbf{r}) + k_\nu^2\mathbf{E}'(\mathbf{r}) = -4\pi N(\nu/c)^2\mathbf{P}_\nu^{(K)}\exp(-i\mathbf{k}\cdot\mathbf{r}), \qquad (3.6)$$

where k_ν is the wave number of the polarization wave,

$$k_\nu = \frac{n_\nu\nu}{c}, \qquad (3.7)$$

and $k_\nu \neq k$ in general.

Aside from Eq. (3.6), the amplitude $\mathbf{E}'(\mathbf{r})$ must satisfy the second Maxwell equation

$$\operatorname{div}\mathbf{E}'(\mathbf{r}) = 0. \qquad (3.8)$$

The solution to Eq. (3.6) is the sum of the general solution of the appropriate homogeneous equation and a particular solution of the inhomogeneous equation. Hence, we have

$$\mathbf{E}'(\mathbf{r}) = \mathbf{A}\exp(i\mathbf{k}_\nu\cdot\mathbf{r}) + \mathbf{B}\exp(-i\mathbf{k}_\nu\cdot\mathbf{r}) + \mathbf{C}\exp(-i\mathbf{k}\cdot\mathbf{r}), \quad (3.9)$$

where the length of the vector $\mathbf{k}_\nu$ is determined by (3.7), and the coefficient $\mathbf{C}$ in the particular solution is given by the following formula:

$$\mathbf{C} = 4\pi N\left(\frac{\nu}{c}\right)^2\left(k^2 - k_\nu^2\right)^{-1}\mathbf{P}_\nu^{(K)}. \qquad (3.10)$$

The fact that the polarization wave is transverse—the condition (3.8)—implies that according to (3.9) the polarization vector $\mathbf{E}'$ is perpendicular to the wave vector $\mathbf{k}_\nu$ of the polarization wave and to the vector $\mathbf{k}$.

To determine the constant vectors $\mathbf{A}$ and $\mathbf{B}$ we must specify the boundary condition at the frequency ν of the polarization wave. The common statement of the problem consists in specifying an interface between the nonlinear medium and a vacuum. The external electromagnetic waves are incident on the nonlinear medium from the vacuum (and are also reflected by the medium). The boundary condition for a field with frequency ν is that $\mathbf{E}'(\mathbf{r})$ at the interface is zero, that is, there must be no polarization wave at the interface between the two media. For the sake of simplicity we will assume that, first, the interface is a plane, and, second, this plane is perpendicular to $\mathbf{k}$. This corresponds to the common experimental setting.

Let us orient the z axis along the vector $\mathbf{k}$, with the interface corresponding to the plane $z = 0$. The nonlinear medium occupies the half space

$z > 0$. The transversality condition (3.8) is met if we assume that $\mathbf{E}'(\mathbf{r})$ lies in the xy plane. From (3.9) we see that the condition $\mathbf{E}'(x, y, z = 0) = 0$ is satisfied only if $\mathbf{k}_\nu$ has no components along the x and y axes. Thus, both $\mathbf{k}_\nu$ and $\mathbf{k}$ are directed along the z axis, which is perpendicular to the interface. This means that the polarization wave is directed along the incident wave. Then solution (3.9) assumes a simpler form:

$$\mathbf{E}'(z) = \mathbf{A} \exp(ik_\nu z) + \mathbf{B} \exp(-ik_\nu z) + \mathbf{C} \exp(-ikz), \quad (3.11)$$

that is, the vectors $\mathbf{k}$ and $\mathbf{k}_\nu$ can be replaced by the corresponding scalars. We see that the amplitude of the polarization wave depends only on the z coordinate in the direction of its propagation.

The next step is to put $\mathbf{A} = 0$, since the corresponding nonlinear reflected wave, whose field strength is proportional to $\exp(i\nu t + ik_\nu z)$, is weak compared to $\mathbf{E}'$ (compare the similar statement for linear optics in Chapter 1). Nonlinear reflection will be discussed in Section 3.5.

Combining (3.11) with the boundary condition $\mathbf{E}'(z = 0) = 0$, we find that $\mathbf{B} = -\mathbf{C}$. Thus, the final expression for the amplitude of the polarization wave is

$$\mathbf{E}'(z) = 4\pi N \left(\frac{\nu}{c}\right)^2 \left(k^2 - k_\nu^2\right)^{-1} \mathbf{P}_\nu^{(K)} \left[\exp(-ikz) - \exp(-ik_\nu z)\right]. \quad (3.12)$$

We see that the intensity of the polarization wave is proportional to the square of density N of the gaseous medium.

3.1.4. The Phase-Matching Condition

From the experimental point of view it is natural to try to create the polarization waves of highest intensity for given intensities of the incident waves. As the general solution (3.12) shows, the closer the values of k and k_ν the greater the intensity. The maximum amplitude of the polarization wave is obtained when

$$k = k_\nu. \quad (3.13)$$

This is known as the *phase-matching condition*. If it is met, then (3.12) yields

$$\mathbf{E}'(z) = -2\pi i N \left(\frac{\nu}{c}\right) z \exp(-ik_\nu z) \mathbf{P}_\nu^{(K)}. \quad (3.14)$$

We see that the field strength in the polarization wave grows linearly with the z coordinate. Of course, this solution is valid only at distances z for which E' remains low compared to the field strengths E_i of the incident waves, so that the unperturbed-field approximation holds true.

Often there is no way in which the exact equality in (3.13) can be achieved (as we will show using concrete examples). We introduce the difference $\Delta \mathbf{k} = \mathbf{k}_\nu - \mathbf{k}$, which is known as the *phase detuning*, and assume that the phase-matching condition can be met at least approximately, that is,

$$\Delta k \ll k_\nu \approx \frac{\nu}{c}. \tag{3.15}$$

If this is so, we can simplify the expression (3.12) for the field strength in the polarization wave:

$$\mathbf{E}'(z) = 2\pi N \frac{\nu}{c\,\Delta k} \mathbf{P}_\nu^{(K)} \left[1 - \exp(i\,\Delta k\,z) \right] \exp(-ik_\nu z). \tag{3.16}$$

We see that, strictly speaking, only when the condition (3.15) [including (3.13)] is met can we isolate in the solution a wave factor that characterizes a plane monochromatic wave,

$$\mathbf{E}'(z, t) = \mathbf{E}_\nu(z) \exp(i\nu t - ik_\nu z), \tag{3.17}$$

and the amplitude of the field strength in the polarization wave that slowly varies with z:

$$\mathbf{E}_\nu(z) = 2\pi N \frac{k_\nu}{\Delta k} \left[1 - \exp(i\,\Delta k\,z) \right] \mathbf{P}_\nu^{(K)}. \tag{3.18}$$

Thus, only when the phase-matching condition (3.15) is met approximately can we speak of the electromagnetic polarization wave as a wave whose amplitude varies slowly from point to point.

Let us now introduce the quantity $L_{\text{coh}} = (\Delta k)^{-1}$ known as the *coherence length*. From solution (3.18) we see that the amplitude of the field strength increases over the coherence length L_{coh}. Indeed, as long as z is much smaller than L_{coh}, the factor $[1 - \exp(i\,\Delta k\,z)]$ in (3.18) increases and attains a maximum at $z \sim L_{\text{coh}}$. If we increase the z coordinate still further, the amplitude in the polarization wave drops (again to zero), then grows, etc. The period of oscillations of $\mathbf{E}_\nu$ over z is of the order of L_{coh}.

The maximum value of the amplitude of the field strength is given by the following formula:

$$E_\nu^{\text{max}} = 4\pi N k_\nu L_{\text{coh}} P_\nu^{(K)}. \tag{3.19}$$

We see that it is directly proportional to the coherence length L_{coh}. When L_{coh} is finite, the applicability condition for the unperturbed-field approximation, $E_\nu^{max} \ll E_i$, as (3.19) shows, is independent of z, in contrast to the case considered above when $L_{coh} = \infty$ ($\Delta k = 0$).

How phase matching can be realized in practice will be shown later in concrete physical examples. Here we wish only to note two more facts of principal significance.

First, not in all nonlinear effects does the condition of phase matching reduce to all wave vectors pointing in the same direction. When several waves interact and the waves propagate in the medium in different directions, the phase-matching condition is realized exactly only for a certain angle between the wave vectors of the propagating waves. Such processes will be considered in Section 3.4.

Second, one must bear in mind that phase matching is not always necessary for an incident wave of one frequency propagating in a medium to generate a wave with another frequency. As an example we cite stimulated Raman scattering (Section 3.3), when the scattered light of a frequency that differs from that of the incident light is a wave propagating in the direction of the incident wave in the particular, but practically important, case where the incident wave is spatially limited in the direction perpendicular to the vector **k** (Section 3.3.2). The phase-matching condition (3.13) is then satisfied identically. Note also that when many waves are interacting there are processes that require phase matching only for some of the interacting waves (Section 3.4).

Processes that take place in the interaction of waves when phase matching is required, that is, processes obeying (3.16) and occurring over the coherence length L_{coh}, are known as coherent processes (Akhmanov and Khokhlov, 1966), in contrast to incoherent processes, which do not require phase matching (see the end of the Introduction to this book). Incoherent processes occur, as we will subsequently see, over the absorption length of the incident radiation. We use this terminology below. Note that sometimes more detailed definitions of coherent and incoherent processes are used, definitions that allow for the nature of the exciting wave (coherent or incoherent) and for the coincidence (or difference) of the initial and final states of the atom in whose spectrum the transitions occur (see the end of the Introduction to this book). One must also bear in mind, that, as has just been said, dividing processes into coherent and incoherent is to some extent arbitrary, since there are processes of mixed type.

3.1.5. Conclusion

If we compare the linear polarization wave (Chapter 1) and the nonlinear polarization wave, we can conclude that in the nonlinear case, in contrast to the linear, either exact or approximate phase matching is required for effective excitation of a polarization wave. In the linear case, obviously, the phase-matching condition is satisfied identically. The direction of propagation of the polarization wave coincides with that of the exciting wave. This is an important conclusion, from which, for one thing, it follows that as the incident wave propagates in the medium, a situation may arise in which the process of interaction of the medium with two waves, the incident wave and the polarization wave, must be considered. This phenomenon underlies, for instance, the method of exciting higher harmonics in stimulated Raman scattering of light (Section 3.3). As for the amplitude of the nonlinear polarization wave, in general it is small compared to the amplitude of the linear polarization wave. It may become large, as (3.19) implies, only when there are resonances in the nonlinear susceptibility, at large intensities of the exciting radiation and a large coherence length, and in a dense medium.

3.2. HIGHER OPTICAL HARMONICS

3.2.1. Introduction

Excitation of higher optical harmonics in a nonlinear medium is one of the most typical and at the same time simple processes. The simplicity of this process lies in the fact that the process occurs when there is only one wave of frequency ω incident on the nonlinear medium. Depending on the properties of the medium, the field strength of the incident wave, and the conditions for the realization of interaction in the nonlinear medium, waves of multiple frequencies $K\omega$ can appear. In reality only waves with frequencies up to 9ω have been observed, but this value is not the limit, since it is determined solely by the experimental technique. Excitation of higher optical harmonics is a typical process because phase matching is required, that is, this is a coherent process.

As already noted in Chapter 2, in the media we are discussing here (specifically, in atomic media), only odd harmonics can be generated. For this reason we will focus on the process of excitation of the lowest odd-numbered harmonic, the third. But we will start our discussion with the excitation of the second harmonic, as the simplest case.

Note that the excitation of optical harmonics in gases constitutes a nonlinear-optics phenomenon that is very important from the practical standpoint, since it is the most useful method yet of obtaining intense coherent radiation in the ultraviolet frequency range.

3.2.2. Classification of the Basic Processes

We start by classifying the basic processes of harmonic excitation. As already noted above, the excitation of even-numbered harmonics in atomic gases in the absence of other fields is impossible because of the law of parity conservation for atomic states. This is a rigorous statement in the dipole approximation. In principle, excitation of even-numbered harmonics in systems possessing central symmetry is possible due to higher multipole moments: the magnetic dipole moment and the electric quadrupole moment. Interaction with higher multipole moments becomes possible when additional strong external fields are employed, such as a constant electric or magnetic field. However, these phenomena do not play an important role and will not be studied in detail here.

Thus, the simplest excitation process in the presence of one external field is the excitation of the third harmonic. The corresponding nonlinear susceptibility $\chi^{(3)}(3\omega; \omega, \omega, \omega)$, which determines the nonlinear polarization, was considered in detail in Section 2.1.

Because the external field is low compared to the characteristic atomic field, the nonlinear polarization $\mathbf{P}^{(3)}$ is small, generally speaking. At a fixed value of the external field strength, $\chi^{(3)}$ and the effectiveness of the excitation of the harmonic increase sharply when there are one or more resonances with the intermediate atomic states.

The excitation of the fifth, seventh,... harmonics is determined, respectively, by $\chi^{(5)}, \chi^{(7)},\ldots$. In the absence of resonances, the polarization $P^{(2K+1)} \propto \chi^{(2K+1)}E^{2K+1}$ drops rapidly as K increases.

In principle, in addition to these processes there may be other processes in which not only are photons of the external field absorbed, but other

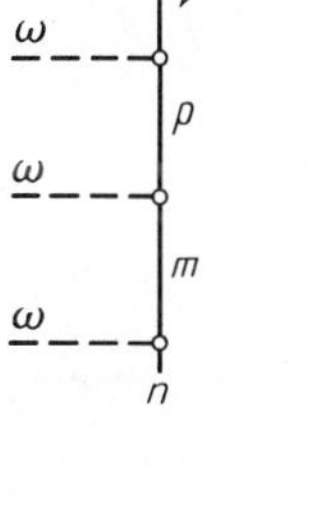

Figure 3.1. The Feynman diagram for third-harmonic excitation accompanied by reemission of a photon of frequency ω'.

photons are emitted. An example is the process for which the Feynman graph corresponding to $\chi^{(5)}$ is depicted in Figure 3.1. This is also a process in which the third harmonic is excited, but it proceeds thanks not only to absorption of photons from the external field but to emission of photons as well (that is, reemission). Compared to the process of third-harmonic excitation due solely to the absorption of photons from the external fields, this latter process is higher in order (in the number of absorbed photons). In what follows we will not consider processes with reemission, because of their low probability in the nonresonance case as compared to higher-order excitation processes without reemission. The fact that reemission processes have such low probability is explained by the higher values of K associated with such processes. Only in the resonance case can such processes become important (Section 3.2.6).

3.2.3. The Second Harmonic of the Incident Radiation

We start with the simplest case, where one monochromatic electromagnetic wave is incident on the nonlinear medium. The electric field strength in this wave is $\mathbf{E} = \mathbf{E}_\omega \exp(i\omega t - ik_\omega z)$, where

$$k_\omega = \frac{n_\omega \omega}{c} \qquad (3.20)$$

is the wave number of the incident wave.

As noted in Section 3.2.1, we wish to start with quadratic polarization. In accordance with (3.2), it can be written in the form

$$\mathbf{P}^{(2)} = \chi^{(2)}(2\omega; \omega, \omega)\mathbf{E}_\omega^2 \exp(2i\omega t - 2ik_\omega z). \qquad (3.21)$$

Here $\chi^{(2)}$ is the second-order nonlinear susceptibility, with the typical diagram depicted in Figure 3.2. As for its tensor structure, the simplest situation emerges if we assume $\mathbf{E}$ to be linearly polarized. Denoting the polarization axis by x (the reader will recall that the z axis points in the direction of propagation of the incident wave), we find that the only

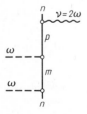

Figure 3.2. The Feynman diagram for second-harmonic excitation.

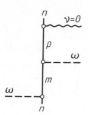

Figure 3.3. The Feynman diagram for constant-electric-field excitation.

component that is nonzero is $\chi^{(2)}_{xxx}$, which corresponds to conservation of the magnetic quantum numbers of the states in the virtual transitions $n \to m \to p \to n$. The vector $\mathbf{P}^{(2)}$ given by (3.21) is parallel to the polarization vector of the incident wave, or simply $\mathbf{E}$.

The expression (3.21) represents only the part of the quadratic polarization that corresponds to the double frequency, but there is also a constant part corresponding to the difference frequency, or the zero frequency:

$$\mathbf{P}'^{(2)} = \chi^{(2)}(0; \omega, -\omega)\mathbf{E}^2_\omega. \qquad (3.22)$$

The typical diagram for $\chi^{(2)}(0; \omega, -\omega)$ is depicted in Figure 3.3. The nonlinear polarization (3.22) generates a constant electric field in the medium. The phase-matching condition (3.13) is satisfied identically here, with the result that the process is incoherent.

On the basis of the general results obtained in Section 3.1.3 we can conclude that the quadratic polarization (3.21) excites a weak electromagnetic field with a frequency 2ω, that is, the second harmonic of the monochromatic incident radiation. The direction of propagation of this second harmonic coincides with that of the incident wave (the z axis). As for the polarization of the harmonic, in the case of linear polarization of the incident radiation the harmonic is polarized in the same direction, while for elliptically polarized incident radiation the nonlinear susceptibility tensor $\chi^{(2)}_{ijk}$ leads in general to a polarization of the harmonic that differs from that of the incident radiation (Akhmanov and Khokhlov, 1966). A detailed discussion of quadratic polarization and second-harmonic excitation can be found in Armstrong, Bloembergen, Ducing, and Pershan, 1962.

On the basis of the general result (3.17) we arrive at the following expression for the electric field strength in the second harmonic:

$$\mathbf{E}'(z, t) = \mathbf{E}_{2\omega}\exp(2i\omega t - 2ik_{2\omega}z), \qquad (3.23)$$

where, in accordance with (3.18), the amplitude of the field is

$$\mathbf{E}_{2\omega}(z) = 2\pi N\chi^{(2)}\mathbf{E}^2_\omega\frac{k_{2\omega}}{\Delta k}[1 - \exp(i\,\Delta k\,z)], \qquad (3.24)$$

and the wave number is

$$k_{2\omega} = \frac{n_{2\omega}2\omega}{c}.$$ (3.25)

The phase detuning Δk has the form

$$\Delta k = k_{2\omega} - 2k_\omega = (n_{2\omega} - n_\omega)\frac{2\omega}{c}.$$ (3.26)

Let us rewrite (3.24) in the equivalent form

$$E_{2\omega}(z) = -8\pi iN\chi^{(2)}E_\omega^2\frac{k_\omega}{\Delta k}\sin\frac{\Delta k z}{2}\exp\left(\frac{i\Delta k z}{2}\right).$$ (3.27)

Note that from (3.27) follows the typical nonlinear (in this case quadratic) dependence of the electric field strength at the second-harmonic frequency on the electric field strength of the incident radiation: $E_{2\omega}(z) \propto E_\omega^2$.

The above expression holds true if the phase-matching condition (3.15) is satisfied; the condition (3.26) implies that phase matching is realized if $|n_{2\omega} - n_\omega| \ll n_\omega$, that is, $|n_{2\omega} - n_\omega| \ll 1$ for a gaseous medium.

From the solution (3.27) follows the criterion of weakness of the second-harmonic wave, $E_{2\omega} \ll E_\omega$, that is, the criterion of applicability of the unperturbed-field approximation for the incident wave of frequency ω:

$$N\chi^{(2)}E_\omega \ll |n_{2\omega} - n_\omega| \ll 1.$$ (3.28)

As we see, this criterion requires moderate intensities of the incident radiation, a low density of the gas, and limits from above and below on the dispersion of the gaseous medium. The upper bound is realized in the neighborhood of the resonant values of the refractive index, while the lower bound is determined by the coherence length [see Eqs. (3.37)–(3.38)].

3.2.4. The Third Harmonic

By means of simple modifications we can generalize the results obtained in Section 3.2.3 for exciting the second harmonic to the case of third-harmonic excitation. By analogy with (3.21), we write the following expression for the cubic atomic polarization:

$$P^{(3)} = \chi^{(3)}(3\omega; \omega, \omega, \omega)E_\omega^3\exp(3i\omega t - 3ik_\omega z),$$ (3.29)

where $\chi^{(3)}$ is the third-order nonlinear atomic susceptibility (Section 2.1). If for the sake of simplicity we assume the incident wave to be linearly polarized (with x the polarization axis and z the direction of propagation of the wave), then $\mathbf{P}^{(3)}$ in (3.29) is directed along the x axis, and $\chi^{(3)}$ is understood to be the $\chi^{(3)}_{xxxx}$ component of the susceptibility tensor.

In accordance with the general results obtained in Section 3.1, the nonlinear polarization (3.29) leads to the appearance of a polarization wave with a frequency 3ω, the frequency of the third harmonic. By simple modifications of (3.24) we arrive at the following expression for the electric field strength in the third harmonic (in the unperturbed-field approximation):

$$\mathbf{E}_{3\omega}(z) = 2\pi N\chi^{(3)}\mathbf{E}_\omega^3 \frac{k_{3\omega}}{\Delta k}[1 - \exp(i\,\Delta k\,z)], \qquad (3.30)$$

or

$$\mathbf{E}_{3\omega}(z) = -12\pi i N\chi^{(3)}\mathbf{E}_\omega^3 \frac{k_\omega}{\Delta k}\sin\frac{\Delta k\,z}{2}\exp\left(\frac{i\,\Delta k\,z}{2}\right), \qquad (3.31)$$

which is similar to (3.27). Here

$$k_{3\omega} = n_{3\omega}\frac{3\omega}{c} \qquad (3.32)$$

is the wave number, and

$$\Delta k = (n_{3\omega} - n_\omega)\frac{3\omega}{c} \qquad (3.33)$$

is the phase detuning [similar to (3.26)]. The quantities n_ω and $n_{3\omega}$ are the refractive indices at the frequency of the incident radiation and at the frequency of the third harmonic, respectively. As (3.33) implies, the lower the dispersion of the refractive index, that is, the weaker the dependence of n on ω, the better is the phase matching and the greater is the amplitude (3.31) of the polarization wave.

The flux of the energy of the polarization wave (3.31) per unit time across a unit cross-sectional area of the medium, that is, the intensity of the polarization wave, is determined by the Poynting vector averaged over time:

$$I_{3\omega} = c|\mathbf{E}_{3\omega}|^2/2\pi. \qquad (3.34)$$

Substituting (3.31) into (3.34), we find that

$$I_{3\omega} = 576\pi^4\left(\frac{k_\omega}{\Delta k\,c}\right)^2(N\chi^{(3)})^2 I_\omega^3\sin^2(\Delta k\,z). \qquad (3.35)$$

Here we have introduced the notation

$$I_\omega = \frac{c|\mathbf{E}_\omega|^2}{2\pi},\qquad\qquad (3.36)$$

which is the mean intensity of the incident wave. The dimensionless ratio $I_{3\omega}/I_\omega$, known as the *transformation coefficient*, characterizes the transformation of the incident radiation into the harmonic. Note that the intensity $I_{3\omega}$ is proportional to the square of the gas density, N^2, the cube of the intensity of the incident radiation, I_ω^3, and the square of the absolute value of the nonlinear susceptibility, $|\chi^{(3)}|^2$ (the nonlinear susceptibility may in general contain a real and imaginary part). The general expressions and the numerical values for $\chi^{(3)}$ were given in Section 2.1, where we saw that $\chi^{(3)}$ varies rapidly with ω, and becomes infinite when the differences in atomic spectral terms, ω_{mn}, coincide with ω, 2ω, or 3ω. In the interresonance regions $\chi^{(3)}$ may vanish, as does the intensity of the third harmonic.

If we increase ω, there will be a frequency at which three-photon ionization of the atom becomes energetically possible. This happens at $3\omega > \mathscr{E}_n$. Such a process results in a nonresonance three-photon absorption of photons from the external electromagnetic field and is obviously in competition (Section 3.2.6) with the third-harmonic excitation process considered here. Mathematically this corresponds to an imaginary part appearing in $\chi^{(3)}$.

If $3\omega > \mathscr{E}_n$, then by varying ω we can achieve a three-photon resonance with an autoionization state. In this case $\chi^{(3)}$ increases (Armstrong and Wynne, 1977; Geller and Popov, 1981), but this also leads to a rise in the probability of multiphoton ionization of the atom (Section 3.2.6).

As we have seen in Section 2.1, variation of ω may also lead to an intermediate one-, two-, or three-photon resonance in the atomic spectrum. This obviously leads to an increase in $\chi^{(3)}$ in comparison to the nonresonance case. However, another consequence is the rise in the probability of competitive processes connected with resonance absorption of the incident radiation (Section 3.2.6).

3.2.5. Phase Matching in a Gas

As already noted in Section 3.1.4, for a sizable portion of the energy of the incident radiation to be transformed into the energy of the wave of another frequency via nonlinear interaction, the phases of these waves must be matched over the maximum length of their propagation in the medium. The phase-matching condition (3.15) is always met over a certain length of the path of the wave [the coherence length $L_{coh} = (\Delta k)^{-1}$], but, in accordance with (3.33), for an arbitrary frequency this length proves to be extremely small, due to the dispersion in the refractive index of the medium. For instance, if Δk is of the order of k_ω, then $L_{coh} \sim \lambda_\omega \sim 10^{-5}$ cm. From (3.33) we see that the phase-matching condition in the case of third-harmonic excitation is reduced to $n_{3\omega}$ being approximately equal to n_ω, where n_ω is the refractive index of the medium at ω. According to (3.33),

$$L_{coh} = c[3(n_\omega - n_{3\omega})\omega]^{-1}. \qquad (3.37)$$

This relationship readily yields the following estimate: for L_{coh} to be about 1 cm, with ω lying in the visible range, the indices of refraction must be equal to high accuracy:

$$\Delta n = n_\omega - n_{3\omega} \sim 10^{-5}. \qquad (3.38)$$

In the absence of phase matching we have $\Delta n = n_\omega - n_{3\omega} \sim 2\pi N\chi^{(1)} \sim 10^{-3}$ for atomic gases [e.g. see (3.42)]. Thus, we must increase the degree of approximation to which n_ω and $n_{3\omega}$ are equal by at least two orders of magnitude.

Various methods have been suggested and implemented that enable realizing the phase-matching condition over a considerable length (3.37) of propagation of the exciting radiation. In the case of gaseous media, the method of increasing the coherence length (3.37) involves choosing an appropriate mixture of vapors of alkali metals with noble gases, where the concentration of the metal vapors is small compared with that of the noble gases. This is known as the *buffer-gas method*.

Before studying the essence of the method, let us discuss the necessary properties of the linear refraction indices of the components in the mixture. These properties follow from the results obtained in Section 1.3.4. The refractive index is connected with the dielectric constant of the gas in a simple manner, namely, $n_\omega = \epsilon_\omega^{1/2}$, while ϵ_ω is related to the linear atomic susceptibility $\chi^{(1)}$ thus (see Section 1.3.4):

$$\epsilon_\omega = 1 + 4\pi N\chi^{(1)}(\omega; \omega). \qquad (3.39)$$

In accordance with the results obtained in Section 1.1 (the incident wave is assumed to be linearly polarized along the x axis),

$$\chi^{(1)}(\omega; \omega) = \sum_m |x_{mn}|^2 \left[(\omega_{mn} - \omega)^{-1} + (\omega_{mn} + \omega)^{-1}\right], \qquad (3.40)$$

where n is the ground state of an atom in the gas being considered. In the static case, $\omega = 0$, the dc susceptibility $\chi^{(1)}(0; 0)$ is positive [as follows from (3.40)] and hence n_0 is greater than unity. Actually n_0 differs from unity very little, because the number of atoms per unit volume of the gas, N, is low (see Section 1.3).

In the case of noble gases the first excited levels m lie very high ($\mathscr{E}_m > 10$ eV), so that in the optical range up to ultraviolet frequencies the size of ω in (3.40) is small compared to ω_{mn}. Expanding (3.40) in a Taylor series in powers of ω, we find that the following formula holds true with a high accuracy (in the case of noble gases):

$$n_\omega = n_0 + n'\omega^2, \tag{3.41}$$

where n_0 is related to the dc susceptibility of the gas via

$$\begin{aligned} n_0 &= 1 + 2\pi N\chi^{(1)}(0; 0) \\ &= 1 + 4\pi N \sum_{m \neq n} |x_{mn}|^2\omega_{mn}^{-1} > 1, \end{aligned} \tag{3.42}$$

while

$$n' = 8\pi N \sum_{m \neq n} |x_{mn}|^2\omega_{mn}^{-3} > 0. \tag{3.43}$$

Since $n' > 0$ and $n_0 > 1$, according to (3.41) n_ω is greater than unity and grows with ω (in this case one speaks of normal dispersion of the refractive index).

In the case of alkali atoms, the energy of the lowest excited state m differs relatively little from the energy of the ground state n ($\mathscr{E}_m \sim 1-2$ eV; see Figure 3.4). For ω lower than ω_{mn}, n_ω is greater than unity, as in the case of noble gases. In the vicinity of the first excited state m, the susceptibility (3.40) increases resonantly, changing its sign at $\omega = \omega_{mn}$. Hence, for sufficiently high frequencies $\omega > \omega_{mn}$ we find that $\chi^{(1)}(\omega; \omega)$ is negative and $n_\omega < 1$. The dispersion is *anomalous*.

The buffer noble gas has a marked effect on the phase-matching condition $n_{3\omega} \approx n_\omega$, since for a noble gas under normal dispersion n_ω and $n_{3\omega}$ are both greater than unity, while for an alkali gas we have $n_\omega > 1$ and $n_{3\omega} < 1$ for some values of ω, due to anomalous dispersion. Thus, by choosing the composition of the mixture it proves possible to compensate, at frequency 3ω, the anomalous dispersion of the alkali vapor with the normal dispersion of the buffer noble gas (Figure 3.5).

As for the nonlinear atomic susceptibility $\chi^{(3)}$, it is much larger for alkali atoms at a frequency ω in the visible range than for atoms of noble gases, due to the small denominators in the formula (2.15) for $\chi^{(3)}$ in the case of

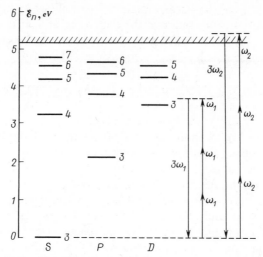

Figure 3.4. The diagram of low-lying excited states of a sodium atom and the diagram of third-harmonic excitation by the radiation from an Nd-glass laser ($\omega_1 \approx 1.17$ eV) and that of a ruby laser ($\omega_2 \approx 1.7$ eV). The numbers on the levels are the principal quantum numbers.

alkali atoms. Thus, the excitation of the third harmonic is caused by the nonlinear polarization of the alkali atoms rather than the noble-gas atoms.

In accordance with the formula (3.39) for an alkali vapor and (3.41) for a buffer gas, the refractive indices of a gaseous mixture at frequencies ω and 3ω have the form

$$n_\omega = n_{0i} + n'_i\omega^2 + 2\pi N\chi_a^{(1)}(\omega; \omega), \tag{3.44}$$

$$n_{3\omega} = n_{0i} + n'_i 9\omega^2 + 2\pi N\chi_a^{(1)}(3\omega; 3\omega), \tag{3.45}$$

where N is the number of alkali atoms per unit volume of vapor, $\chi_a^{(1)}(\omega; \omega)$

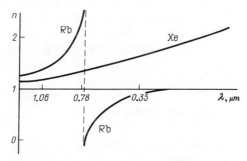

Figure 3.5. Refractive indices of an alkali vapor and a buffer noble gas as functions of the radiation frequency ω (wavelength λ).

and $\chi_a^{(1)}(3\omega; 3\omega)$ are the linear atomic susceptibilities of the alkali vapor, n_{0i} is the static refractive index of the buffer gas, and n_i' is determined by the formula (3.43) for a buffer gas. Hence, in the buffer-gas method the condition of phase matching $n_{3\omega} = n_\omega$ assumes the form

$$32\omega^2 N_i \sum_m |x_{mn}|^2 \omega_{mn}^{-3} = N\left[\chi_a^{(1)}(\omega; \omega) - \chi_a^{(1)}(3\omega; 3\omega)\right]. \quad (3.46)$$

According to (3.43), N_i is the number of atoms of the buffer gas per unit volume. In ordinary experimental conditions N is usually much less than N_i. This fact is compensated in (3.46) by the large (negative) value of $\chi_a^{(1)}(3\omega; 3\omega)$ and the fact that for a noble gas the ratio $(\omega/\omega_{mn})^2$ is small.

Experiments have shown (see Doitcheva, Mitev, Pavlov, and Stamenov, 1978; Jong, Bjorklund, Kung, Miles, and Harris, 1971; and Ohashi, Ishibashi, Kobayasi, and Inaba, 1976) that the accuracy of calculations and methods for mixing gases and vapors with given ratios of components is quite high for the practical use of the buffer-gas method. Typical mixtures have a total pressure of the order of 10^2 to 10^3 Torr and a relative partial pressure of the alkali vapors ($\sim N/N_i$) of the order of 10^{-5} to 10^{-2}. The design of the vessels in which the vapors and gases mix is complicated (Bloom, Bekkers, Young, and Harris, 1975; Puell, Spanner, Falkenstein, Kaiser, and Vidal, 1976), since high homogeneity of the mixture is required in a sufficiently large volume (the linear dimensions are several meters) and at a definite pressure, which determines the pressure of the alkali-atom vapors.

3.2.6. Limiting and Competing Effects

In this section we will study the physical phenomena that restrict the possibility of obtaining large coefficients of transformation of the incident radiation into the harmonics. As Eq. (3.35) shows, there are three ways in which the transformation coefficient can be increased: (1) by increasing the number N of atoms of the process gas, that is, by increasing the pressure of the gas (the transformation coefficient is proportional to N^2), (2) by increasing the intensity I_ω of the incident wave (the transformation coefficient into the Kth harmonic is proportional to I_ω^K), and (3) by achieving resonance with an intermediate atomic level (this leads to a resonant increase in the nonlinear susceptibility $\chi^{(K)}$, and the transformation coefficient is proportional to $|\chi^{(K)}|^2$). From what we have just said, the most promising seems to be the second method, especially at high values of K.

A number of phenomena that occur in the process of excitation of harmonics in a gas may hinder the increase of the transformation coefficient. These are the one- and multiphoton absorption of the incident radiation and the multiphoton ionization of atoms. The role of these effects increases when there appears an intermediate resonance between the frequency of the external field and a transition in the atomic spectrum. Other important effects here are the change in the nonlinear susceptibility brought on by the external field, and also the optical breakdown of the process gas. We will study these effects separately and establish their role in relation to the parameters that characterize the external field and the atomic gas.

3.2.6.1. One-Photon Absorption. (See Section 1.4.) Incident photons may be resonantly absorbed by the gas, which leads to transition of the gas atoms into excited states. The excited states may then decay in a cascade, with emission of spontaneous photons having frequencies and directions that generally differ from the frequency and direction of propagation of the incident wave. Thus, in one-photon absorption both the number of the incident photons and the number of atoms in the initial state decrease. These two factors diminish the effectiveness of the transformation of incident radiation into harmonics, so that one-photon absorption constitutes a competitive process. Note that from the practical standpoint the decrease of the number of photons plays a smaller role, because, as is well known (see Hanna, Yuratich, and Cotter, 1979), the number of photons passing through a cell with a gas is several orders of magnitude greater than the number of atoms in the cell. It is obvious then that resonance absorption is the highest when there is an exact resonance within the transition width in the atomic spectrum and the spectral width of the exciting radiation. The spectral line shape of resonance absorption for a pressure of the buffer gas in the cell that is of practical interest (several tenths of an atmosphere) is determined by collisions of the alkali atoms with the atoms of the buffer gas rather than by the Doppler effect or, a fortiori, the natural linewidth of the excited state or the spectral width of the exciting radiation.

Obviously, the role of one-photon collisional absorption grows with the number of atoms of the atomic mixture per unit volume, that is, with the pressure in the atomic mixture. The reader will recall that it is the alkali atoms that do the main absorption, since in view of the smallness of the energies of their lowest excited states, the resonance misfits $|\omega_{mn} - \omega|$ are much smaller than in the buffer noble gas, for which the energies of the lowest excited states lie much higher.

The dominant role of one-photon absorption arises from the fact that this is a first-order process in the field strength of the electromagnetic wave, while harmonic excitation is a Kth-order process. This implies that one-photon absorption constitutes the main competing effect that hinders attempts to increase $\chi^{(K)}$ by creating a one-photon resonance with a discrete atomic level, although, as Eq. (2.15) shows, $\chi^{(K)}$ does increase.

In the strong field of the incident wave there occurs one-photon resonant mixing of the ground level and an excited level (the saturation effect; Section 1.2). Here a strong field is understood to be a field for which $x_{mn}E_\omega > \max[\omega_{mn} - \omega, \Gamma_m]$, where Γ_m is the resonance width. If the field strength is increased, the absorption first increases and then becomes saturated, since the populations of the ground (n) and excited (m) levels become equal (Section 1.2).

Experiments have enabled determining (see Jong, Bjorklund, Kung, Miles, and Harris, 1971, and Puell, Spanner, Falkenstein, Kaiser, and Vidal, 1976) the energy flux of the exciting radiation at which one-photon absorption manifests itself. This value proved to be an order of magnitude (or even more) greater than the calculated value (see Sobelman, Vainstein, and Yukov, 1981); in the calculations the resonance width was determined from the theory of collision broadening. Possibly this theory cannot be applied to the case of large resonance misfits, which were inherent in the experiments, and one must use the quasistatic theory of collision broadening, which is valid for regions lying very far from the center of the absorption line (Vidal and Cooper, 1969).

In the weak field of the incident wave, when no resonance mixing of levels occurs (Section 1.2), the role of absorption depends on the relationship between the duration of the exciting wave, T, and the natural lifetime of an excited level, τ, which lifetime determines the probability of decay of this level. It is clear that when T is much longer than τ, the absorption is noticeably higher than when T is much shorter than τ, since the absorption probability is proportional to T. Thus, here the use of short pulses of exciting radiation is preferable if the transformation coefficient is to be maximal. Note that since in practice the resonance width is much greater than the natural width of the level (which we have already spoken of), a weak field (in accordance with the above criterion) may be strong enough for exciting harmonics in a gaseous medium.

3.2.6.2. Two-Photon Absorption. In two-photon absorption (for details see Section 3.7), an atom absorbs two photons and goes over to an excited state p. The atom may then decay with the emission of several spontaneous photons, which generally have frequencies and directions of emission that differ from those of the photons from the external field. As in one-photon absorption, photons leave the incident beam and the number of atoms in the ground state decreases, which means that this process may be considered competitive with harmonic excitation. In contrast to one-photon

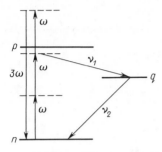

Figure 3.6. Two-photon absorption accompanied by the deexcitation of two photons with frequencies ν_1 and ν_2, a process competitive with third-harmonic excitation.

absorption, which can play an important role even at low intensities of the excited radiation, two-photon absorption manifests itself only at rather high intensities. The schematic of the two-photon absorption process is shown in Figure 3.6.

In the absence of a two-photon resonance, two-photon absorption is low. However, as noted above, it is advisable to achieve such a resonance with a view to increasing the nonlinear susceptibility $\chi^{(3)}$ and thus increasing the probability of third-harmonic excitation. This also increases the probability of two-photon absorption as a competing process. Note that the intensity of the harmonic (3.35) and the probability of two-photon absorption are proportional to Δ^{-2}, where $\Delta = |\omega_{pn} - 2\omega|$ is the misfit (detuning) of the two-photon resonance.

In a strong field (or when the tuning to resonance is good, that is, $\Delta = 0$), the probability of two-photon excitation of level p in unit time is described by (2.45), which in the given case takes the form

$$w_{pn}^{(2)} = 2\left|\chi_{pn}^{(1)}\right|^2 |E_\omega|^4 \Gamma_p^{-1}, \qquad (3.47)$$

where $\chi_{pn}^{(1)}$ is the linear susceptibility (the off-diagonal two-photon compound matrix element), E_ω the field strength in the incident wave, and Γ_p the total width of level p. The saturation of two-photon absorption occurs when

$$w_{pn}^{(2)} t > 1, \qquad (3.48)$$

where $t = \min(T, \Gamma_p^{-1})$, with T the duration of the external field. Note that in a strong light field the resonance transition width Γ_p depends on the field strength E_ω and may exceed the collisional width of level p (and therefore be considerably greater than the natural width τ^{-1} of level p). When

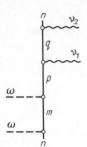

Figure 3.7. The Feynman diagram for the nonlinear susceptibility describing two-photon absorption (see Figure 3.6).

saturation occurs, the probability of two-photon absorption remains unchanged even if the field strength in the external wave is increased.

Two-photon absorption accompanied by emission of spontaneous photons corresponds to a nonlinear susceptibility determined by the Feynman diagram in Figure 3.7. In a sufficiently strong field there may occur, in addition to this process, another process associated with stimulated emission. This stimulated emission of photons is depicted in Figure 3.8, and the nonlinear susceptibility corresponding to the process has the following form (see Section 2.4):

$$\chi^{(3)}(\omega; \omega, \omega, -\omega) = \left(\omega_{pn} - 2\omega + i\Gamma_p\right)^{-1}\left[\sum_m x_{nm}x_{mp}\left(\omega_{mn} - \omega\right)^{-1}\right]^2,$$

(3.49)

that is, the nonlinear susceptibility increases resonantly as the misfit of the two-photon resonance decreases. Puell, Scheingraber, and Vidal, 1980, were the first to note the possibility of this last process proving highly competitive in the presence of a two-photon resonance.

As for the misfit of a two-photon resonance, $\omega_{pn} - 2\omega$, we must note that in contrast to the similar quantity for a one-photon resonance, the

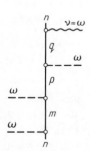

Figure 3.8. The Feynman diagram for the nonlinear susceptibility describing two-photon absorption accompanied by stimulated emission of two photons of the same frequency.

former must be estimated and calculated with allowance made for the quadratic Stark shifts of the energies of states p and n in an alternating field (Section 1.1.4), which are of the same order of magnitude in the field strength in the incident wave as the field broadening in two-photon absorption (precisely, proportional to E_ω^2). The relatively large size of this shift disrupts two-photon resonances quite easily.

Another undesirable effect that occurs in two-photon absorption in saturation is that by diminishing the population of the ground state n the saturation also changes the common linear nonresonance susceptibility $\chi_{nn}^{(1)}$. This in turn leads to a change in the linear refractive index, which is connected with $\chi_{nn}^{(1)}$ via (3.44)–(3.45). The result is an uncontrolled change in the phase, detuning Δk, and hence the phase-matching condition is violated (Section 3.2.5).

In a number of concrete cases (the type of interaction, the radiation frequency, the composition of the gas mixture) a restriction on the intensity of harmonics caused by two-photon absorption has been observed (Kildal and Brueck, 1980; Puell, Scheingraber, and Vidal, 1980; Vidal and Cooper, 1969).

3.2.6.3. Multiphoton Ionization of Atoms.

This constitutes a competitive effect in the exciting of harmonics in gases. The competitiveness of the process is quite obvious: a number of atoms are ionized, and hence the initial number of atoms with which the interaction occurs diminishes; also, the number of photons in the incident electromagnetic wave diminishes, but this usually plays no practical role, since the number of photons is considerably greater than the number of atoms.

There is a marked difference between the process of harmonic excitation and that of multiphoton ionization from the point of view of the concentration of atoms, N. The multiphoton-ionization probability is proportional to N (incoherent process), while (see above) the harmonic-excitation probability is proportional to N^2 (coherent process). For this reason an increase in the density of the medium, N, diminishes the competitiveness of multiphoton ionization (Agarwal and Tewari, 1983; Tewari, 1983). Since the probability of multiphoton ionization is proportional to the duration T of the incident radiation, we can diminish the competitiveness of ionization by diminishing this duration. As for the resonant increase in the effectiveness of the third-harmonic excitation process, the presence of a two-photon resonance with an intermediate level increases the probability of multiphoton resonance ionization.

The role of multiphoton ionization as a competitive process is clearly seen in the results obtained by Miyazaki and Kashiwagi, 1978, who observed third-harmonic excitation after applying the radiation of a YAG

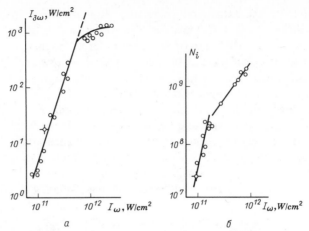

Figure 3.9. (*a*) Third-harmonic intensity and (*b*) multiphoton-ionization yield as functions of the intensity of the incident radiation (Miyazaki and Kashiwagi, 1978).

laser ($\omega = 9445$ cm^{-1}) to a mixture of sodium and xenon with total pressure of about 1 Torr. The experimenters observed a simultaneous release of radiation at 3ω and Na$^+$ ions created as a result of direct five-photon ionization (at the above-noted frequency this is the lowest-order process allowed by energy conservation). Figure 3.9 illustrates the experimental data on the dependence of the third-harmonic yield on the incident intensity. As can be clearly seen , at $I_\omega = (3\text{–}5) \times 10^{11}$ W/cm^2 the intensity of the third harmonic starts to increase more slowly than by the $I_{3\omega} \propto I_\omega^3$ law. This effect is due to five-photon ionization, which leads to noticeable ionization of the atoms of the gaseous medium at approximately the same radiation intensity ($N_i \propto I_\omega^5$).

In this case, too, the appearance of resonances leads to an increase in the probability of both processes occurring: the main process of higher-harmonic excitation and the competitive process of multiphoton ionization. With regard to intermediate resonances, their role is qualitatively similar to that of the described competitive process of atom excitation. A new situation emerges when a resonance state is the final state in the excitation of a harmonic ($\omega_{qn} \approx 3\omega$). In this case the role of resonances is clearly seen from the results of experiments involving excitation of the third harmonic with laser radiation passing through mercury vapor (Normand, Morellec, and Reif, 1983). By changing the laser frequency the experimenters achieved resonances with different bound electronic states. The schematic of the excitation is shown in Figure 3.10. When the laser frequency ω_1 was tuned

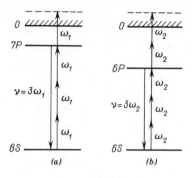

Figure 3.10. Third-harmonic excitation in mercury vapor accompanied by (*a*) one-photon ionization and (*b*) two-photon ionization following the resonance absorption of three photons from the incident radiation.

to the three-photon 6*S*-7*P* resonance, the transition from the 7*P* state to the continuous spectrum took place as a one-photon process, and its probability was so high that the ionization process completely suppressed the third-harmonic excitation process. When a three-photon resonance at ω_2 was achieved on the 6*S*-6*P* transition, the transition from the resonance 6*P* state to the continuous spectrum was already two-photon. The probability of two-photon ionization from a resonance state is much lower than the probability of one-photon ionization, with the result that what was observed was third-harmonic excitation, which successfully competed in this case with resonance ionization.

One must bear in mind that multiphoton ionization of atoms may influence the excitation of harmonics not only directly but also by forming a partially ionized plasma and hence changing the refractive index of the medium. This leads to disruption of phase matching and decrease in the harmonic yield.

Up till now we have spoken of multiphoton excitation of an atomic state in the discrete spectrum. Let us now turn to the case where an autoionization state in the continuous spectrum of an atom undergoes a multiphoton resonant excitation. Alber and Zoller, 1983, found that tuning to such a resonance decreases the competitiveness of multiphoton excitation. On the one hand, as we approach resonance, the third-harmonic yield increases, since it is proportional to the square of the cubic susceptibility, $|\chi^{(3)}|^2$, and hence in resonance is inversely proportional to the square of the autoionization-state width, Γ_p^{-2}, since $\chi^{(3)} \propto \Gamma_p^{-1}$; but on the other, the probability of multiphoton ionization of the ground state of an atom in autoionization resonance is proportional to Γ_p^{-1}, that is, lower than that of third-harmonic excitation. This ensures the decrease in competitiveness of multiphoton ionization in resonance.

Dimov, Pavlov, Geller, and Popov, 1983, and Pavlov, Dimov, Metchkov, Mileva, and Stamenov, 1982, have studied the effect that autoionization-like

resonances have on the competitiveness of multiphoton ionization in relation to harmonic excitation. These resonances exist in the continuous
spectrum of every atom, are induced by an external electromagnetic field,
and constitute quasi energy states with energies $\mathscr{E}_n + K\hbar\omega$ lying in the
continuous spectrum. If a second field is superimposed on this system with
frequency equal to that of the transition from such an autoionization state
to an excited state in the discrete spectrum, then the relative yield of the
third harmonic increases in comparison with the multiphoton-ionization
probability, just as it did in the previous case. There are two reasons for this
increase. The first is related to the direct nonlinear process in which three
photons of the incident radiation of frequency ω are absorbed and the atom
passes from the ground state into an autoionization-like state of the
continuous spectrum. Next there occurs a resonant emission and absorption
of a photon of the second field of frequency ω', and after this a third-
harmonic photon is emitted and the atom returns to its ground state (Figure
3.1). This process is described by the nonlinear susceptibility
$\chi^{(5)}(3\omega; \omega, \omega, \omega, \omega', -\omega')$. The second mechanism of the increase in the
third-harmonic yield is associated with the addition of the frequency 3ω of
the third harmonic, which is formed by the cubic polarization
$\chi^{(3)}(3\omega; \omega, \omega, \omega)$ from the incident radiation of frequency ω, to ω'. This
cascade mechanism (for details see Section 3.2.9) in the second stage is
described by the cubic susceptibility $\chi^{(3)}(3\omega; 3\omega, \omega', -\omega')$.

These processes have been realized in the experiments of Dimov, Pavlov,
Geller, and Popov, 1983, and Pavlov, Dimov, Metchkov, Mileva, and
Stamenov, 1982, who excited the third harmonic in sodium vapor by
employing the radiation of a 530-nm dye laser. It was found that after
absorbing three photons, an atom went over from the ground state $3S$ to
the autoionization-like state in the continuous spectrum with an energy
$\mathscr{E}(3S) + 3\hbar\omega$. When the second field tuned to the one-photon resonance
with the transition from the above-noted autoionization-like state to the
excited state $5S$ was turned on, it was found that the intensity of excitation
of the third harmonic increased five to ten times.

3.2.6.4. Nonlinear Variation of the Refractive Index. According to (1.51),
the linear refractive index n_ω is related to the linear susceptibility of the
medium at frequency ω through the following formula:

$$n_\omega = 1 + 2\pi N\chi^{(1)}(\omega; \omega), \qquad (3.50)$$

which implies that it is independent of the field strength. A dependence
emerges when we allow for the next terms in the series expansion of the
susceptibility in powers of the field strength. In accordance with the

definition (2.1), we must replace $\chi^{(1)}(\omega; \omega)$ in (3.50) with

$$\chi^{(1)}(\omega; \omega) + \chi^{(3)}(\omega; \omega, -\omega, \omega)\mathbf{E}_\omega^2. \qquad (3.51)$$

The cubic susceptibility $\chi^{(3)}(\omega; \omega, -\omega, \omega)$ is determined via (2.17). Figure 2.5 shows the Feynman diagrams for this quantity.

Thus, instead of (3.50), in a strong field there emerges a refractive index of the medium, $n_\omega(E_\omega)$, which depends on the field strength of the exciting wave:

$$n_\omega(E_\omega) = n_\omega + n_\omega^{(2)}E_\omega^2, \qquad (3.52)$$

where we have introduced the notation

$$n_\omega^{(2)} = 2\pi N\chi^{(3)}(\omega; \omega, -\omega, \omega). \qquad (3.53)$$

The change in the medium's refractive index given by (3.52) can be accounted for in the phase-matching condition. Of course, in doing so allowance must be made for the change in the refractive indices of the alkali vapors and the buffer gas.

One must bear in mind, however, that the refractive index is determined not only by properties of the alkali vapors and the buffer gas but also by properties of the plasma of free electrons formed in the process of multiphoton ionization of the gas atoms. If the field strength in the incident wave is high, the number of electrons per unit volume, $N_e(E_\omega)$, may be considerable. The corresponding addition to the refractive index can be found in an elementary manner by calculating the energy of free electrons in the field of a monochromatic wave. In atomic units we have

$$n_\omega^{(e)} = -\frac{2\pi N_e}{\omega^2}. \qquad (3.54)$$

This expression is written for the case of a linearly polarized field. In a strong field this contribution must also be taken into account in the phase-matching condition.

3.2.6.5. Optical Breakdown of the Gas Mixture by Laser Radiation. The physical phenomena that cause optical breakdown of gases are well known (e.g. see Raĭzer, 1977). The criterion for such breakdown to occur has the following form (Rapoport, Zon, and Manakov, 1978, Chapter 5):

$$NTE_\omega^2 > 10^{23}, \qquad (3.55)$$

where N (cm^{-3}) is the number of gas atoms per unit volume, $T(s)$ the duration of the radiation, and E_ω (V/cm) the amplitude of the field strength of the radiation causing the breakdown. The size of the constant on the right-hand side of (3.55) corresponds to the optical frequency range we are interested in, to the absence of radiation focusing (or to weak focusing), and to the nanosecond range of pulse length T. From (3.55) it follows that for the highest possible pressure of the gas mixture (of the order of 100 Torr) and the greatest possible length of the laser pulse with Q switching ($T \sim 10^{-7}$ s), the field strength E_ω is of the order of 10^6 V/cm, which corresponds to intensities I_ω of the order of 1 GW/cm^2.

In some cases optical breakdown may limit the effectiveness of transformation of laser radiation into higher harmonics obtained through high field strengths E_ω of the exciting radiation. Besides, from (3.55) we see that the intensity of the breakdown field $I_\omega \propto E_\omega^2$ is inversely proportional to the medium's density N. For this reason breakdown may limit the extent to which N may be increased in order to increase the intensity of the harmonics (the reader will recall that $I_{K\omega} \propto N^2$).

3.2.7. Excitation of the Second Harmonic in Gases in the Presence of a Constant Electric Field

As repeatedly noted earlier, excitation of the second harmonic in atomic gases by a monochromatic light field is impossible because it would require violation of parity conservation. However, in the presence of a constant electric field such a process becomes possible. One typical Feynman diagram for the cubic susceptibility describing this process is depicted in Figure 3.11. Corresponding to this diagram is the following quantity:

$$\chi^{(3)}_{ijkl}(2\omega; \omega, \omega, 0) = \sum_{mpq} r^i_{nm} r^j_{mp} r^k_{pq} r^l_{qn}$$

$$\times \left[(\omega_{mn} - \omega)(\omega_{pn} - 2\omega)(\omega_{qn} - 2\omega) \right]^{-1} + \cdots . \quad (3.56)$$

Note that because of the large number of possible permutations of the

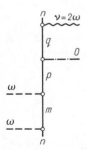

Figure 3.11. One of the Feynman diagrams for the cubic susceptibility corresponding to second-harmonic excitation in the presence of a constant electric field.

order in which the various fields act, there are many more Feynman diagrams in this case than for third-harmonic excitation.

The intensity of the second harmonic, $I_{2\omega}$, has the form

$$I_{2\omega} \propto N^2 \left(\chi^{(3)} I_\omega E_c \right)^2, \qquad (3.57)$$

where I_ω is the intensity of the incident electromagnetic wave, and E_c the strength of the constant electric field. According to (3.15), phase matching is achieved if

$$\Delta k = \left(n_{2\omega} - n_\omega \right) \frac{2\omega}{c} \ll k. \qquad (3.58)$$

As for the tensor structure of $\chi^{(3)}$, in contrast to the case of third-harmonic excitation by a single field, where only one component of the susceptibility tensor was nonzero, here there are several nonzero components, depending on the mutual orientation of the constant electric field strength and the polarization vector of the incident electromagnetic wave. The work of Finn and Ward, 1971, provides an example of such a process. Note that the incident laser radiation may play the role of the second radiation if the field amplitude is distributed inhomogeneously over the wavefront (Mossberg, Flusberg, and Hartman, 1978).

One must bear in mind that from the standpoint of applications the excitation of the second harmonic in gases in the presence of a constant electric field is of considerable interest, since its realization requires an incident electromagnetic wave with an intensity lower than required for third-harmonic excitation. But the strength of the constant electric field must be at least no lower than that of the varying electromagnetic field.

3.2.8. Excitation of the Second Harmonic in Gases in the Presence of a Constant Magnetic Field

Excitation of the second harmonic in atomic gases becomes possible via the selection rules if the spontaneously emitted photon is not a dipole photon but an electric quadrupole ($E2$) photon. Of course, the probability of an $E2$ photon being emitted is much lower than that of an $E1$ photon. The corresponding factor is $a^2/\lambda^2 \ll 1$, where a is the size of the atom, and λ is the wavelength of the radiation. But this is compensated by the fact that in the given case we are dealing with quadratic nonlinearity, rather than cubic as in the above case.

Bearing in mind all that has been said, we can write

$$\chi^{(2)}(2\omega; \omega, \omega) \propto x_{nm} x_{mp} \left[x(\mathbf{k}_{2\omega} \cdot \mathbf{r}) \right]_{pn}. \qquad (3.59)$$

Here it is assumed (just as earlier) that the incident electromagnetic wave is linearly polarized along the x axis and propagates along the z axis. Then the states n, m, and p have the same magnetic quantum numbers. The quantity $\mathbf{k}_{2\omega} \cdot \mathbf{r}$ in (3.59) is the additional factor for the spontaneously emitted electric quadrupole photon.

The quadrupole matrix element in (3.59) implies that $\mathbf{k}_{2\omega} \cdot \mathbf{r} = k_{2\omega}x$, that is, $\mathbf{k}_{2\omega}$ is directed along the x axis, which is perpendicular to the vector $\mathbf{k}_{\omega}$ of the incident electromagnetic wave. Thus, there is no coherent polarization wave of frequency 2ω that propagates in the same direction as the incident wave, as there is with third-harmonic excitation. We have thus established that no excitation of the second harmonic occurs, although parity conservation allows for such excitation.

To lift the forbiddenness with respect to the magnetic quantum number and obtain nonzero matrix elements in (3.59), a constant magnetic field must be applied in the direction of the y axis, which is perpendicular to the x axis (along which the electric field in the incident electromagnetic wave is directed) and to the z axis (along which vector $\mathbf{k}_{\omega}$ is directed). This magnetic field mixes the atomic states in relation to the magnetic quantum numbers, as a result of which the quadrupole matrix element $[xzk_{2\omega}]_{pn}$ in (3.59) becomes nonzero, since $\mathbf{k}_{2\omega} \cdot \mathbf{r} = k_{2\omega}z$. Thus, $\mathbf{k}_{2\omega}$ is now directed along the z axis (just as $\mathbf{k}_{\omega}$ is), and the emerging polarization wave propagates into the medium rather than in a transverse direction (as it did in the previous case). We can conclude, therefore, that the second harmonic of the incident electromagnetic wave is excited (Bethune, Smith, and Shen, 1978).

When the magnetic field is weak, mixing in the magnetic quantum numbers occurs in the first order of perturbation theory and hence is linear in the magnetic field strength H_c. The intensity of the second harmonic, being proportional to $|\chi^{(2)}|^2$, therefore proves to be proportional to the square of the magnetic field strength, H_c^2.

We can write the following expression for the intensity of the second harmonic:

$$I_{2\omega} \propto N^2 \left(\chi^{(2)} I_\omega H_c \right)^2. \tag{3.60}$$

If this is compared with (3.57), we see that the main difference is that in a constant electric field the second-harmonic intensity is determined by the cubic susceptibility $\chi^{(3)}(2\omega; \omega, \omega, 0)$, while in a constant magnetic field it is determined by the quadratic susceptibility $\chi^{(2)}(2\omega; \omega, \omega)$.

Dunn, 1983, experimentally realized the above scheme by exciting the second harmonic in sodium vapor. The wavelength of the incident electromagnetic wave, 578.7 nm, was such that the two-photon resonance of the ground state $3S$ of the sodium atom and the excited state $5S$ was realized. The transverse (constant) magnetic field was 175 G. It was found that the intensity of the excited second harmonic is, in accordance with the theoretical result (3.60), proportional to the square of the gas density, N^2. The resonance width was determined by the Doppler effect.

3.2.9. Excitation of Higher Harmonics in Gases

The excitation of higher harmonics in gases is of obvious interest as one of the promising methods for obtaining coherent radiation in the ultraviolet range. If we are interested in linear absorption in gases, that is, absorption caused by one-photon transitions, then it must be noted that linear transparency extends to the vacuum ultraviolet. Indeed, to excite atoms of noble gases into the first excited states, photon energies of about 8 eV for Xe to 19 eV for He are required, while for ionization these values are 12 to 24 eV, respectively. However, we must bear in mind that the excitation of higher harmonics is caused by higher-order terms in the expansion of the nonlinear polarization in powers of the field strength, which are extremely small themselves. Therefore, for effective transformation of the incident radiation into higher harmonics a high strength of the excitation radiation field is essential, and we must allow for the nonlinear transparency of the medium. We are speaking of the same processes of multiphoton excitation and multiphoton ionization of the atoms of a medium as those considered in Section 3.2.6 as limiting and competing effects in the excitation of the third harmonic. In the case of higher harmonics it is clear that the number of concrete processes of nonlinear absorption will be greater than in the case of third-harmonic excitation.

A new phenomenon that has an analog in the excitation of the third harmonic is the presence of a channel of cascade excitation of higher harmonics, in addition to the channel of direct excitation. A direct process is one of the kind we have just considered, that is, the absorption by an atom of an odd number of photons of frequency ω that leads to emission of a photon of frequency $K\omega$ equal to the sum of frequencies of the absorbed photons. This process is characterized by the nonlinear susceptibility $\chi^{(K)}$, where K is the corresponding odd number. In contrast to this, a cascade process consists of two or more stages: first the fixed frequency $K\omega$ is obtained as a result of a new radiation of frequency $K'\omega$ appearing (where $K' < K$), and then there is an interaction with the medium of two (or several) waves with frequencies ω and $K\omega$ (or other multiples of ω). This process is determined by the nonlinear susceptibility $\chi^{(K')}$ and other susceptibilities, depending on the concrete scheme of the second stage. The yield of the Kth-harmonic radiation is determined by the sum of the nonlinear susceptibilities of the direct $(\chi_d^{(K)})$ and cascade $(\chi_{cas}^{(K)})$ processes. For instance, in the excitation of the fifth harmonic both the direct process $\omega + \omega + \omega + \omega + \omega = 5\omega$ and the cascade process $\omega + \omega + \omega = 3\omega$, $3\omega + \omega + \omega = 5\omega$ contribute. As an example we take the ratio of the nonlinear susceptibilities for these processes in sodium vapor (Miyazaki, Sato, and Kashiwagi, 1979): $\chi_d^{(5)}/\chi_{cas}^{(5)} \approx 2$. Note that the calculation of $\chi_{cas}^{(K)}$ poses no additional problems in comparison with the calculation of $\chi_d^{(K)}$, while separating a direct process from a cascade requires special experiments (Miyazaki, Sato, and Kashiwagi, 1979).

3.2.10. The Practical Realization of Higher-Harmonic Excitation in Gases

From the practical standpoint, the excitation of harmonics in gases is of great interest as an effective method of generating intense coherent radiation in the ultraviolet range. Three main parameters characterize harmonic excitation from the standpoint of application: the intensity of the excited radiation, the transformation coefficient of the incident radiation into the harmonic, and the frequency (or wavelength) of the excited radiation. Optimization of these parameters requires a much more complicated technique for realizing the harmonic-excitation process in gases than under the conditions discussed above.

For instance, if we wish to obtain extremely shortwave radiation, we need to realize the excitation of higher harmonics, either direct or cascade. Hence the interest in higher nonlinearities in the expansion of the polarization vector in powers of the field strength.

The need to increase the intensity of the exciting radiation and the transformation coefficient to a higher harmonic changes the conditions of transformation considerably from those discussed above.

First, the transformation occurs in conditions that violate the unperturbed-field approximation. The qualitative difference of this case from the one discussed above lies in the saturation of the harmonic yield, that is, in the deviation from the $I_{K\omega} \propto E_\omega^{2K}$ law, where E_ω is the strength of the exciting field, toward slower growth of the intensity. The fact is that when the energy is transferred from the incident light wave to the harmonic, the incident wave weakens, that is, E_ω decreases. The theoretical description of the process grows more complicated, since we must solve nonlinear Maxwell equations. Already in the unperturbed-field approximation we saw that after partial energy transfer occurs over the coherence length L_{coh} from the incident light wave to the harmonic, the next process is the inverse transfer of energy from the harmonic to the incident wave. Such alternation of energy transfer to the harmonic and back along the z axis in the medium takes place outside the framework of the unperturbed-field approximation.

Energy transfer may be considerable (with the total energy of the electromagnetic field remaining the same, obviously) and may even be complete.

Second, it proves to be expedient to focus the incident radiation onto the gas cell so as to increase the strength of the exciting field. This poses the problem of optimizing the phase matching of focused beams. The principal difference of this process from the one discussed above is that optimal transformation is achieved at $\Delta k \neq 0$. A theoretical description of such a process and the expression for the dependence of the optimal size of Δk on the parameters that characterize the condition for the focusing of laser radiation have been discussed in Bjorklund, 1975; Tomov and Richardson, 1976; and Ward and New, 1969.

Third, and last, laser radiation is always only quasimonochromatic. The nonmonochromaticity of laser radiation is caused by the complicated mode composition of the radiation and the finiteness of the pulse length in pulsed lasing. The nonmonochromaticity of the incident radiation, characterized by $\Delta\omega$, increases the phase detuning Δk.

There is no need to list quantitative data on the parameters characterizing the transformation of laser radiation into harmonics in gases. These are technical problems, which are solved faster than books are written and published. For a general picture we only note that the radiation of a solid-state pulsed Nd-glass laser ($\lambda \sim 1\ \mu$m) is transformed into harmonics up to the ninth (Kung, Jong, and Harris, 1973), while the shortest wavelength available in 1984 was $\lambda = 38$ nm, which was obtained in the form of the seventh harmonic of the fourth harmonic of the Nd-glass laser radiation (Reintjes, She, and Eckardt, 1978). A complete list of all the data on harmonic excitation obtained up to 1979 for various gases, mixtures, and frequencies of the exciting radiation is given in Hanna, Yuratich, and Cotter, 1979, Section 4.5.

3.2.11. Excitation of Harmonics in Other Media

Another possibility lies in the excitation of even-numbered harmonics of the incident radiation in noncentrosymmetric crystals. The simplest such process is the excitation of the second harmonic.

Crystals differ from gases in two respects. First, the linear transparency of crystals is limited to wavelengths of about 0.3 μm, that is, the near ultraviolet region. For this reason from the practical standpoint the most interesting transformation into the harmonics of the radiation of high-power lasers in the visible range is limited to the second (or sometimes the third) harmonic of the incident radiation. Second, the only possibility of ensuring phase matching of the waves in crystals is to use anisotropic crystals and to match the phase of different waves, at least one of which must be an extraordinary wave. This drastically limits the class of crystals that can be used to excite harmonics and makes additional demands on the exciting radiation (in particular its polarization) and on the orientation of the crystal with respect to the wave vector and the polarization vector of the radiation.

Practically all these problems have already been solved (Dmitriev and Tarasov, 1982; Zernike and Midwinter, 1973), and crystal transducers of visible laser radiation into the second harmonic are available commercially.

If several crystals oriented in a certain manner are placed in the beam of the incident radiation, higher harmonics can be obtained through the cascade process, which emerges as a result of the interaction of two waves: the incident wave and a harmonic. For instance, the third harmonic of the radiation of an Nd-glass laser is obtained in precisely this manner by employing two crystals. The linear transparency of crystals, as said earlier, limits such a process to the near-ultraviolet region of the spectrum.

Finally, we must note that in using crystals there appear strong limits on the power of the incident radiation compared to the case of harmonic excitation in gases. This is due to the lower threshold of optical breakdown in crystals.

3.2.12. Conclusion

The process of harmonic excitation described in Section 3.2 is a typical nonlinear-optics process, which has no analog in linear optics. Indeed, within the scope of linear optics, only spontaneous Raman scattering changes the frequency of the light that interacts with the medium. Although, in principle, the frequency of the light may increase in spontaneous Raman scattering by an arbitrary amount, this requires anti-Stokes scattering and, hence, a medium consisting of excited atoms, that is, an unstable target. The increase in the light frequency accompanying excitation of harmonics, as we saw earlier, can be realized for any initial state of the atoms of the medium, including the ground state.

If we use the language of averaged optical characteristics of a medium (which is common practice), then harmonics are excited as a result of the nonlinear polarization of the medium. But the essence of the excitation process is seen most clearly if we turn to the elementary nonlinear-optics phenomena and employ the microscopic language. At the atomic level the difference lies in the fact that within the scope of linear optics the absorption of light is limited to one-photon processes, while in nonlinear optics multiphoton absorption of light plays the dominant role. Thus, at the atomic level the specific feature of the excitation of harmonics lies in the multiphoton nature of the absorption of light by an atom and hence in the multiphoton excitation of the atom.

If we now turn to the macroscopic characteristics of the process of harmonic excitation, the specific feature is that the transfer of a sizable part of the energy from the exciting wave into the polarization wave requires that the phase-matching condition for these waves be satisfied. This, in turn, requires the excited wave to be spatially and temporally homogeneous. The polarization waves at the frequencies of the higher optical harmonics are in this case coherent, a situation that differs drastically from the case of the incoherent light produced in a medium by spontaneous Raman scattering.

In conclusion we note once more that the specific features of the excitation of higher harmonics, at the base of which lies multiphoton absorption of the light incident on the medium and the emergence of nonlinear polarization of this medium, clearly points to the fact that this process can take place only when the field strength in the incident light wave is sufficiently high.

3.3. STIMULATED RAMAN SCATTERING

3.3.1. Introduction

In this section we discuss the phenomenon of stimulated Raman scattering. Though also nonlinear, it is more complicated than the process of higher-harmonic excitation discussed above.

At the base of stimulated Raman scattering lies the process of spontaneous Raman scattering resulting from the off-diagonal linear atomic susceptibility $\chi_{np}^{(1)}(\nu; \omega)$ (Section 1.1.5). The reader will recall that in spontaneous Raman scattering the transition from the initial state of an atom, n, to a state m with a higher energy takes place together with absorption of a photon from the external light wave of frequency ω, whereupon the relaxation of state m with transition to the final state p is accompanied by spontaneous emission of a photon with the Raman frequency ν (Figure 1.3). As a rule, n is the ground state of the atom, so that $\nu < \omega$ and Stokes scattering appears (Figure 1.3a).

When the intensity of the incident light wave of frequency ω is high, the intensity of the polarization wave of frequency ν proves to be also high, which leads to an induced transition from state m to state p with a frequency ν. Induced transitions are the reason for the nonlinearity of stimulated Raman scattering, since their probabilities depend on the intensity of the incident light wave, in contrast to the probabilities of spontaneous transitions.

Induced transitions (also called stimulated transitions) with the frequency ν constitute the fundamental component of stimulated Raman scattering.

Thus, in stimulated Raman scattering the incident wave of frequency ω induces a polarization wave of frequency ν in the medium (for the direction of propagation of the latter wave see below). As we will see, emergence of the fundamental component of stimulated Raman scattering has no bearing on the need to satisfy the phase-matching condition.

As soon as in a medium there appears a wave with the Raman frequency ν in addition to the incident wave with frequency ω, the two start to interact. Because of the nonlinearity of the medium, there appear waves with sum and difference frequencies ($2\omega \pm \nu$, and so on), or the so-called higher components of stimulated Raman scattering (or higher Raman frequencies). The interaction in this case is qualitatively the same as in the excitation of harmonics (Section 3.2), namely, the waves of the higher components of stimulated Raman scattering are generated only under phase matching with the waves of frequencies ω and ν.

In describing stimulated Raman scattering we will first turn, as we did in other cases, to the ideal model of interaction, assuming the wave to be monochromatic and the medium to consist of isolated fixed atoms. Later we will study the effects associated with the finiteness of the width of the spectrum of the exciting wave, the thermal motion of the medium atoms, and the collisions of these atoms.

Many writings have been devoted to stimulated Raman scattering, including monographs by Akhmanov and Koroteev, 1981; Dmitriev and Tarasov, 1982; Hanna, Yuratich, and Cotter, 1979; and Lugovoĭ, 1968.

3.3.2. Spontaneous Raman Scattering in an Atomic Medium

In Section 1.1.5 we considered a single atom on which a single spontaneous-Raman-scattering photon is created. Let us now consider a medium consisting of many atoms. As the incident light wave with a frequency ω propagates through the atomic medium, the spontaneous Raman scattering is multiple. We denote (as we did earlier) the direction in which the incident wave propagates by z. The number dn_ν (dimensionless) of spontaneously scattered photons over a length dz (cm) can be expressed in terms of the spontaneous-Raman-scattering cross section $d\sigma_\nu$ (cm^2), the number of photons N_ω (dimensionless) from the incident wave, and the number N of atoms per unit volume (cm^{-3}) thus:

$$dn_\nu = N_\omega N\, d\sigma_\nu\, dz. \tag{3.61}$$

Indeed, $N\, d\sigma_\nu$ (cm^{-1}) proves to be the probability of spontaneous Raman scattering over a length of 1 cm.

By definition, the spontaneous-Raman-scattering cross section $d\sigma$ at frequency $\nu = \omega - \omega_{pn}$ is related to the rate of spontaneous Raman scattering dw_{np} [given by (1.20)] in the following manner:

$$d\sigma = \frac{dw_{np}}{S}, \tag{3.62}$$

where

$$S = \frac{I_\omega}{\omega} = \frac{c|E_\omega|^2}{2\pi\omega} \tag{3.63}$$

is the flux density of the incident photons of frequency ω. Substituting (3.62) into (1.20) yields

$$d\sigma = \omega\nu^3 c^{-4}\left|\chi_{np}^{(1)}(\nu; \omega)\right|^2 d\Omega_\nu. \tag{3.64}$$

This formula gives the cross section for spontaneous Raman scattering accompanied by the emission of a spontaneous photon of frequency $\nu = \omega - \omega_{np}$.

Since state p is excited and has a definite width Γ_p, when the atom goes over from state m to state p, it can emit spontaneous photons with a frequency ν that lies in the neighborhood of $\omega - \omega_{pn}$. To find the probability of spontaneous Raman scattering with emission of a spontaneous photon of frequency ν near $\omega - \omega_{np}$, we must multiply (3.64) by the dimensionless factor $g(\nu)\,d\nu$, where $g(\nu)$ is defined by (1.27) and is the spectral lineshape of the emitted photon (frequency ν). Thus, we have

$$d\sigma_\nu = d\sigma\,g(\nu)\,d\nu = \omega\nu^3 c^{-4}\left|\chi_{np}^{(1)}(\nu; \omega)\right|^2 g(\nu)\,d\nu\,d\Omega_\nu. \tag{3.65}$$

Substituting (3.65) into (3.61) yields

$$dn_\nu = \omega\nu^3 c^{-4}N_\omega N\left|\chi_{np}^{(1)}(\nu; \omega)\right|^2 g(\nu)\,dz\,d\nu\,d\Omega_\nu. \tag{3.66}$$

This expression gives the number of spontaneously emitted photons that are in different states within the following interval of wave vectors: $[\mathbf{k}_\nu, \mathbf{k}_\nu + d\mathbf{k}_\nu]$. The number of such states is

$$V\frac{d\mathbf{k}_\nu}{(2\pi)^3} = V\nu^2\frac{d\nu\,d\Omega_\nu}{(2\pi c)^3}, \tag{3.67}$$

where V is the volume of the system, and $d\Omega_\nu$ is (as above) the differential of the solid angle into which the spontaneous photons are emitted. Dividing (3.66) by (3.67), we find the (dimensionless) number of spontaneous photons per unit state:

$$dN_\nu = (2\pi)^3 \frac{dn_\nu}{V d\mathbf{k}_\nu}$$

$$= (2\pi)^3 \omega\nu(cV)^{-1} N_\omega N \left|\chi_{np}^{(1)}(\nu;\omega)\right|^2 g(\nu)\, dz. \qquad (3.68)$$

3.3.3. Stimulated Photon Emission

Equation (3.68) was obtained on the assumption that the spontaneous photons are created in the absence of photons of the same frequency ν. However, if N_ν spontaneous photons have already been created in a medium, then for a sufficiently large N_ν the process of emission of the $(N_\nu + 1)$st photon will be stimulated and in the right-hand side of (3.68) we must introduce a factor $N_\nu + 1$, with the term $+1$ corresponding to spontaneous emission of a photon, and the term N_ν to the induced (stimulated) emission of a photon with frequency ν. An important aspect of spontaneous emission is that the emitted photon has the same direction of emission and the same polarization as the photons of frequency ν already in the medium.

If the number of photons in the incident light wave with a frequency equal to ω is large, that is, the intensity of the incident wave is high, then as this wave passes through the medium, that is, as z increases, the number N_ν rapidly grows accordingly to (3.68). For $N_\nu \gg 1$ the spontaneous emission can be neglected in comparison with the stimulated emission, and instead of (3.68) we arrive at the following expression for the number dN_ν of photons for stimulated Raman scattering:

$$dN_\nu = G_\nu N_\nu\, dz \qquad (3.69)$$

[which is N_ν times greater than (3.68)]. Here we have introduced

$$G_\nu = (2\pi)^3 \omega\nu(cV)^{-1} N_\omega N \left|\chi_{np}^{(1)}(\nu;\omega)\right|^2 g(\nu), \qquad (3.70)$$

which is known as the *gain*.

Let us express G_ν in terms of the field strength E_ω in the incident wave with frequency ω. To this end we note that the energy of an electromagnetic wave of frequency ω inside volume V can be given in two equivalent forms

$(\hbar = 1)$:

$$N_\omega \omega = \frac{V|E_\omega|^2}{2\pi}. \tag{3.71}$$

Combining (3.71) with (3.70), we arrive at the final expression for the gain:

$$G_\nu = 4\pi \frac{\nu}{c} N \left| \chi_{np}^{(1)}(\nu; \omega) \right|^2 |E_\omega|^2 \Gamma_p \left[(\nu - \omega + \omega_{pn})^2 + \Gamma_p^2 \right]^{-1}. \tag{3.72}$$

Here we have used the explicit expression for the spectral factor $g(\nu)$ from (1.27). Note that, as expected, the arbitrary volume V canceled out of the final expression (3.72). We also see that the gain G_ν increases with the field strength E_ω in the incident light wave.

The maximum value of G_ν is attained at exact resonance at $\nu = \omega - \omega_{pn}$:

$$G_\nu^{\max} = 4\pi N \frac{\nu}{c\Gamma_p} \left| \chi_{np}^{(1)}(\nu; \omega) E_\omega \right|^2. \tag{3.73}$$

Equation (3.72) shows that stimulated Raman scattering is always a resonance process in the sense that $\nu \approx \omega - \omega_{pn}$ (either exactly or approximately). However, by *resonant stimulated Raman scattering* we understand a process in which, in addition to the abovementioned resonance condition, the condition of intermediate one-photon resonance, $\omega \approx \omega_{mn}$, is met. The linear susceptibility $\chi_{np}^{(1)}(\nu; \omega)$, defined in (1.25), grows resonantly in the process. The same can be said of the gain G_ν. When $\nu = \omega - \omega_{pn}$, G_ν also increases in a resonant manner, and its value at $\nu = \omega - \omega_{pn}$ can be obtained by substituting (1.25) into (3.73):

$$G_\nu^{\max} = 4\pi N \frac{\nu}{c\Gamma_p} \left| (\mathbf{r}_{pm} \cdot \mathbf{e})(\mathbf{r}_{mn} \cdot \mathbf{e}_0) E_\omega \right|^2 \left[(\omega_{mn} - \omega)^2 + \Gamma_m^2 \right]^{-1}. \tag{3.74}$$

Here $\mathbf{e}_0$ is the unit polarization vector of the incident wave of frequency ω, and $\mathbf{e}$ is the unit polarization vector of the stimulated-Raman-scattering wave of frequency ν. The expression (3.74), in turn, is at its maximum at $\omega = \omega_{mn}$:

$$G_\nu^{\max} = 4\pi N \frac{\nu}{c\Gamma_p \Gamma_m^2} \left| (\mathbf{r}_{pm} \cdot \mathbf{e})(\mathbf{r}_{mn} \cdot \mathbf{e}_0) E_\omega \right|^2. \tag{3.75}$$

After we have established the properties of the gain in Eq. (3.69), let us turn to the solution of this equation. It has a simple form:

$$N_\nu(z) = N_\nu(0)\exp(G_\nu z),\qquad (3.76)$$

that is, the number of photons of frequency ν grows exponentially with z. Over the path G_ν^{-1} it increases e-fold, so that G_ν^{-1} is the characteristic length on which stimulated Raman scattering occurs. The quantity $N_\nu(0)$ in (3.76) is determined by spontaneous Raman scattering at the interface of the medium and vacuum, that is, at $z = 0$.

3.3.4. Properties of Stimulated Raman Scattering

A feature that clearly distinguishes stimulated Raman scattering from the excitation of harmonics discussed in Section 3.2 is the fact that, as we see from the above, stimulated Raman scattering does not require the condition of phase matching to be imposed on the wave vectors $\mathbf{k}_\omega$ and $\mathbf{k}_\nu$ of the waves with frequencies ω and ν, respectively, so as to ensure effective transformation of radiation with frequency ω into radiation with frequency ν.

Although stimulated Raman scattering is determined by the linear susceptibility $\chi_{np}^{(1)}(\nu; \omega)$, it is an essentially nonlinear process, as is clearly demonstrated by (3.76), which provides an exponential dependence on the electric field strength in the incident light wave. As we will subsequently see, in reality stimulated Raman scattering is determined by the cubic susceptibility $\chi^{(3)}$. However, in view of the resonance condition $\nu \approx \omega - \omega_{pn}$, it can be expressed in terms of the square of the linear susceptibility, $|\chi^{(1)}|^2$ [see Eq. (3.77) below]. This resolves the seeming contradiction by which the linear susceptibility $\chi^{(1)}$ determines a nonlinear process.

We note once more the important difference between stimulated Raman scattering and spontaneous Raman scattering. The results of Section 1.1.5 imply that spontaneous Raman scattering is not a directional process, that is, spontaneous-Raman-scattering photons of frequency ν are emitted in various directions with respect to the direction in which the incident wave propagates (the z axis). On the other hand, in stimulated Raman scattering the emitted Stokes photon of frequency ν has, as noted earlier, the same direction as the absorbed Stokes photon of the stimulated-Raman-scattering wave of frequency ν. Thus, stimulated emission is simply the rescattering of a Stokes photon of frequency ν by the medium. Since spontaneous-Raman-scattering photons propagate in various directions, in principle the excitation of stimulated Raman scattering begins in all directions, too. However, in real experiments a beam of laser radiation is focused on the medium, and the diameter of this beam is small compared to the longitudinal dimension (the length of the cell with the gas) over which the incident wave propagates and stimulated Raman scattering is excited. For this reason the greatest gain is achieved only along the z axis (forward and backward). Thus, in (3.74) the unit vector $\mathbf{e}$ representing the polarization wave of stimulated Raman scattering is directed, as $\mathbf{e}_0$ is, along the x axis (which is perpendicular to the z axis). If we employ the terminology of coherent and incoherent processes given in the Introduction to this book, the nondirectional process of spontaneous Raman scattering provides incoherent radiation, while the directional process of stimulated Raman scattering provides coherent radiation.

In spontaneous Raman scattering the atom returns to the initial state n after absorbing a photon of frequency ω and emitting a Stokes photon of frequency ν by cascade spontaneous deexcitation, in which several photons are emitted. In contrast, in stimulated Raman scattering after a photon of frequency ω has been absorbed and a Stokes photon of frequency ν has been emitted, a Stokes photon of frequency ν is absorbed (this simply means rescattering of the Stokes photon, of which we spoke above), and finally a photon of frequency ω of the incident light wave is emitted. This means rescattering of a photon of the incident wave. Since both N_ν and N_ω are much greater than unity, these processes prove to be much more effective than spontaneous deexcitation of the excited states. Energy conservation in stimulated Raman scattering is satisfied in an obvious manner: $\nu + \omega = \nu + \omega$.

Formally, the above four-photon process can be described by the cubic susceptibility $\chi^{(3)}(\nu; \omega, \nu, -\omega)$. One of the Feynman diagrams for this quantity is shown in Figure 3.12. Owing to the condition $\nu \approx \omega - \omega_{pn}$, the susceptibility is resonantly high and can be represented in the form (2.29), namely,

$$\chi^{(3)}(\nu; \omega, \nu, -\omega) = \left(\omega_{pn} - \omega + \nu + i\Gamma_p\right)^{-1}\left|\chi^{(1)}_{np}(\nu; \omega)\right|^2. \quad (3.77)$$

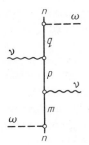

Figure 3.12. One of the Feynman diagrams for the cubic susceptibility corresponding to stimulated Raman scattering for the first Stokes component.

According to Figure 3.12, the expression for the cubic polarization is nonlinear in the field strength E_ω in the incident light wave of frequency ω; precisely, it is quadratic in E_ω.

As Eq. (3.77) shows, the nonlinear susceptibility $\chi^{(3)}$ contains both a real and an imaginary part. Moreover, it is expressed in terms of the off-diagonal linear susceptibility $\chi^{(1)}_{np}(\nu; \omega)$, and this agrees with the fact that above we used this linear susceptibility to describe the gain in stimulated Raman scattering.

Thus, an examination of the Feynman diagrams shows that there is a formal analogy between stimulated Raman scattering and third-harmonic excitation—both processes are determined by cubic susceptibilities. However, while third-harmonic excitation is determined by the cubic susceptibility of the type $\chi^{(3)}(3\omega; \omega, \omega, \omega)$, in stimulated Raman scattering we have to do (see Figure 3.12) with the cubic susceptibility of the type $\chi^{(3)}(\nu; \omega, \nu, -\omega)$. The two processes also differ in the conditions for resonance. In the case of third-harmonic excitation one has to select the frequency ω of the incident light wave in such a manner that there is a two-photon resonance $\omega_{pn} \approx 2\omega$, while in the case of stimulated Raman scattering the frequency ω of the external light field is arbitrary, and the two-photon resonance $\omega_{pn} \approx \omega + \nu$ is established automatically by the stimulated-Raman-scattering frequency ν.

We have thus examined the formation of the fundamental component of the stimulated-Raman-scattering spectrum. We now turn to the mechanism of formation of higher harmonics.

3.3.5.   Raman Frequencies in Stimulated Raman Scattering

After a Stokes wave from stimulated Raman scattering of frequency ν has formed in the medium, it can interact with the initial wave of frequency ω and produce new waves. Here there is a qualitative analogy with the mechanism of formation of higher harmonics from the incident light wave of frequency ω (Section 3.2). The lowest frequencies of the new waves in the case considered here are $2\nu - \omega$ and $2\omega - \nu$ if we bear in mind that according to conservation of parity in atomic states there are no waves with frequencies $\omega \pm \nu$ in a gaseous medium (Sections 3.1 and 3.2). The first of the abovementioned frequencies can be written in the form $\omega - 2(\omega - \nu)$, and, since $\nu < \omega$ for Stokes scattering, it is called the *second Stokes component* of stimulated Raman scattering. This process is called *coherent anti-Stokes Raman scattering*. Note that in stimulated Raman scattering, in contrast to spontaneous Raman scattering, the waves with anti-Stokes frequencies form in a medium consisting of atoms in their ground states because of the interaction of the waves with frequencies ω and ν.

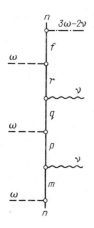

Figure 3.13. The Feynman diagram for the cubic susceptibility describing the excitation of the first anti-Stokes component in stimulated Raman scattering.

If we continue to add frequencies in this manner, we will clearly see that there appear higher anti-Stokes components in the stimulated-Raman-scattering spectrum with frequencies $\omega + K(\omega - \nu)$, $K = 2, 3, \ldots$, together with higher Stokes components in the stimulated-Raman-scattering spectrum with frequencies $\omega - K(\omega - \nu)$, $K = 2, 3, \ldots$. The presence of higher Raman frequencies can be explained by the interaction of the external wave of frequency ω with the wave with the difference frequency $\omega - \nu$, which results in the appearance of waves with frequencies $\omega + K(\omega - \nu)$, where $K = 0, \pm 1, \pm 2, \ldots$. For one thing, the value $K = -1$ corresponds to the first Stokes frequency in the stimulated-Raman-scattering spectrum, ν.

Figure 3.13 gives an example of the Feynman diagram corresponding to the first anti-Stokes component in the stimulated-Raman-scattering spectrum. We see that this diagram is determined by the nonlinear susceptibility $\chi^{(3)}(2\omega - \nu; \omega, -\nu, \omega)$. Owing to the condition $\omega_{pn} \approx \omega - \nu$, this is a resonance susceptibility, just as the susceptibility given by (3.77) was for the fundamental component in the stimulated-Raman-scattering spectrum. The second anti-Stokes component (Figure 3.14) is determined $\chi^{(5)}$ and also

Figure 3.14. The Feynman diagram for the susceptibility $\chi^{(5)}$ describing the excitation of the second anti-Stokes component in stimulated Raman scattering.

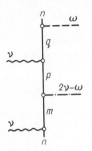

Figure 3.15. The Feynman diagram for the cubic susceptibility describing the excitation of the second Stokes component in stimulated Raman scattering.

represents a resonance susceptibility. As for the second Stokes component, the corresponding susceptibility $\chi^{(3)}(2\nu - \omega; \nu, \nu, -\omega)$ is determined by Feynman diagrams one of which is shown in Figure 3.15. This is again a resonance susceptibility, as seen from Figure 3.15, because of the condition $\omega_{pn} \approx \nu - (2\nu - \omega) = \omega - \nu$, which is necessary for the excitation of the fundamental component with frequency ν in the stimulated-Raman-scattering spectrum. The same condition, obviously, is necessary for exciting all higher components in the stimulated-Raman-scattering spectrum in the interaction of waves with frequencies ω and ν.

As for the anti-Stokes components with the frequency $2\omega + \nu$ and similar components that form as a result of the interaction of the wave whose frequency is ω with the wave having the sum frequency $\omega + \nu$, they are practically not excited, since the sum frequency $\omega + \nu$ lies far from the frequency ω_{pn} of the atomic transition (which is close to $\omega - \nu$) and the corresponding susceptibility does not contain a resonance denominator like the one in (3.77). For instance, for the Feynman diagram depicted in Figure 3.16 and corresponding to the sum anti-Stokes frequency $\omega + (\omega + \nu)$, a similar denominator in $\chi^{(3)}(2\omega + \nu; \omega, \nu, \omega)$ has the form $\omega_{pn} - \omega - \nu$. At $\omega_{pn} \approx \omega - \nu$ it is equal to -2ν and differs considerably from zero, leading to a small value of the susceptibility.

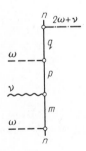

Figure 3.16. The Feynman diagram for the cubic susceptibility describing the excitation of the anti-Stokes component with the sum frequency $2\omega + \nu$.

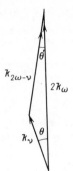

Figure 3.17. Geometrical representation of the phase-matching condition for the excitation of the first anti-Stokes component in stimulated Raman scattering.

A specific feature of the processes of excitation of waves with Raman frequencies in stimulated Raman scattering that distinguishes them from the excitation of the first Stokes frequency ν in the stimulated-Raman-scattering spectrum is the fact they require phase matching. The reason for this is obvious: anti-Stokes and higher Stokes frequencies appear as a result of the interaction of various waves in a medium with dispersion. For instance, for the first Stokes component in the diagram in Figure 3.13, the phase-matching condition takes the form of the following relationship between the wave vectors:

$$2\mathbf{k}_\omega = \mathbf{k}_\nu + \mathbf{k}_{2\omega-\nu}. \tag{3.78}$$

Obviously, for excitation of the first Stokes component (Figure 3.12) the condition has the form $\mathbf{k}_\omega + \mathbf{k}_\nu = \mathbf{k}_\omega + \mathbf{k}_\nu$, that is, the phase-matching condition is satisfied automatically.

The condition (3.78) can be satisfied exactly if we employ a Stokes wave with frequency ν that propagates not exactly along the z axis (that is, in the direction of propagation of the incident wave with frequency ω) but at a certain small angle θ to the z axis. This angle can be found from the diagrammatic representation of (3.78) shown in Figure 3.17. Bearing in mind that the sides of the triangle are $k_\nu = n_\nu\nu/c$, $2k_\omega = 2n_\omega\omega/c$, and $k_{2\omega-\nu} = n_{2\omega-\nu}(2\omega - \nu)/c$, that θ is much less than unity, and that the refractive index in gases is close to unity, we easily arrive at the following expression for this angle:

$$\theta = \left\{ (2\omega - \nu)(\omega\nu)^{-1}\left[n_\nu\nu + (2\omega - \nu)n_{2\omega-\nu} - 2\omega n_\omega \right] \right\}^{1/2}. \tag{3.79}$$

Actually, Eq. (3.79) includes small differences between the refractive indices and unity. The angle θ', at which the first anti-Stokes-component wave propagates, is determined in a similar manner from the triangle in Figure 3.17:

$$\theta' = \frac{\theta \nu}{2\omega - \nu}.$$

$$(3.80)$$

The direction of propagation of other components with the Raman frequencies in the stimulated-Raman-scattering spectrum can be found in a similar manner.

We see that the processes of excitation of the first Stokes and first anti-Stokes components in the stimulated-Raman-scattering spectrum are determined by the nonlinear susceptibility $\chi^{(3)}$, which is of the same order as the susceptibility that leads to the excitation of the fundamental Stokes component with frequency ν in the stimulated-Raman-scattering spectrum. When intermediate resonances are present, the higher-order nonlinear susceptibilities necessary for the excitation of higher Raman frequencies may also prove to be of the same order. Thus, all these processes are competitive with each other and with the excitation of the fundamental Stokes component. Since the intensity of these processes depends essentially on the intensity of the Stokes wave with frequency ν because the higher Raman frequencies are formed from combinations of frequencies ω and ν, the intensity of the higher Stokes components will naturally be high only when the number N_ν of photons with frequency ν is high. In reality this corresponds to saturation (Section 3.3.8), for which $N_\nu \sim N_\omega$. In view of the fact that the field strengths in the waves with frequencies ω and ν are of the same order of magnitude, the fields in the waves with frequencies equal to the Raman frequencies are high. Thus, we can say that there is a threshold in the production of such waves: effective excitation of Raman frequencies is possible only if the intensity of the fundamental Stokes component in the stimulated-Raman-scattering spectrum proves to be of the same order of magnitude as that of the incident wave.

3.3.6. Linear Law of Excitation for Stimulated Raman Scattering

The formula (3.76) for the number of photons produced in stimulated Raman scattering characterizes the increase in intensity of the Stokes radiation in stimulated Raman scattering with frequency ν. More complete information on this radiation is provided by the electric field strength E_ν of stimulated Raman scattering with frequency ν. Indeed, if we characterize the radiation by its intensity, we lose information on the phase of a Stokes wave.

To determine E_ν we must turn, just as we did in Sections 3.1 and 3.2, to the Maxwell equations at frequency ν. In this section we consider the case of small z, when the unperturbed-field approximation is valid at frequencies ω and ν (for a similar approach to determining the higher-harmonic field see Section 3.2). Thus, we assume that on the gaseous medium occupying the half space $z > 0$ there is incident an electromagnetic wave of frequency ω and amplitude E (the decrease of this amplitude in the medium is ignored). In the medium there appears, due to spontaneous radiation, a spontaneous-Raman-scattering electric field of frequency ν and field strength E_ν created by the linear polarization of the medium. This polarization will also be considered fixed. The aim of this section is to calculate the variation δE_ν in E_ν caused by the nonlinear polarization of the medium as a function of z, the distance which the incident wave of frequency ω travels from the interface at $z = 0$. Hence, we restrict our discussion to small z by assuming that $\delta E_\nu \ll E_\nu$.

The cubic polarization (2.7) in the given case has the following form:

$$\mathbf{P}^{(3)}(z) = \chi^{(3)}(\nu; \omega, \nu, -\omega)|\mathbf{E}_\omega|^2\mathbf{E}_\nu\exp(-ik_\nu z), \qquad (3.81)$$

where the nonlinear susceptibility $\chi^{(3)}$ is determined by the Feynman diagram depicted in Figure 3.12 and similar diagrams, and the wave number $k_\nu = n_\nu\nu/c$. The sought quantity $\delta\mathbf{E}_\nu = \delta\mathbf{E}_\nu(z, t)$ is determined by solving the inhomogeneous Maxwell equation (3.4):

$$\frac{\partial^2\,\delta\mathbf{E}_\nu}{\partial z^2} - \left(\frac{n_\nu}{c}\right)^2\frac{\partial^2\,\delta\mathbf{E}_\nu}{\partial t^2} = -4\pi N\left(\frac{\nu}{c}\right)^2\mathbf{P}_\nu^{(3)}(z)\exp(i\nu t) \qquad (3.82)$$

with a known right-hand side. Isolating the time dependence [see (3.5)] in the following manner:

$$\delta\mathbf{E}_\nu(z, t) = \delta\mathbf{E}_\nu(z)\exp(i\nu t), \qquad (3.83)$$

we arrive at the equation [see also (3.6)]

$$\frac{\partial^2\,\delta\mathbf{E}_\nu(z)}{\partial z^2} + k_\nu^2\,\delta\mathbf{E}_\nu(z) = -4\pi N\left(\frac{\nu}{c}\right)^2\chi^{(3)}|\mathbf{E}_\omega|^2\mathbf{E}_\nu\exp(-ik_\nu z) \qquad (3.84)$$

with a given right-hand side. Its solution is obvious [see (3.14)]:

$$\delta\mathbf{E}_\nu(z) = -2\pi iN\frac{\nu}{c}\chi^{(3)}|\mathbf{E}_\omega|^2\mathbf{E}_\nu z\exp(-ik_\nu z). \qquad (3.85)$$

Here we have allowed for the fact that in gases the refractive index n_ν is approximately equal to unity. As in the previous cases, the solution satisfies the boundary condition $\delta E_\nu(z = 0) = 0$. We see that the field $\delta E_\nu(z)$ excited by the nonlinear polarization $\chi^{(3)}(\nu; \omega, \nu, -\omega)$ increases linearly with z. This is the law by which the field strength E_ν increases in stimulated Raman scattering.

If we add the variation $\delta E_\nu(z, t)$ to the field $E_\nu \exp(i\nu t - ik_\nu z)$ in the Stokes wave, then the amplitude of the Stokes wave, E'_ν, is given by the following formula:

$$E'_\nu = E_\nu(1 - if\chi^{(3)}z),\qquad(3.86)$$

where we have introduced the notation

$$f = 2\pi N \frac{\nu}{c}|E_\omega|^2.\qquad(3.87)$$

This yields a formula for the intensity $I' = c|E'_\nu|^2/2\pi$ of the radiation at frequency ν:

$$I'_\nu = I_\nu|1 - if\chi^{(3)}z|^2.\qquad(3.88)$$

Note the essential difference between the intensity I'_ν of the Stokes wave in stimulated Raman scattering and the intensity of the higher harmonics [e.g. see the formula (3.35) for the third harmonic]. This difference is due to the physical fact that in the excitation of harmonics there is no wave at the frequency ν of the harmonic at the boundary, while in the excitation of stimulated Raman scattering there is a spontaneous-Raman-scattering wave at the boundary [the first term on the right-hand side of (3.86)]. For this reason, as (3.35) shows, the intensity of the excited harmonic is proportional (as $z \to 0$) to z^2 and $|\chi^{(3)}|^2$, while from (3.88) we get (by squaring the modulus explicitly)

$$I'_\nu = I_\nu\{1 - ifz(\chi^{(3)} - \chi^{(3)*}) + Az^2\},\qquad(3.89)$$

or, if we restrict our discussion to terms linear in z (for small z),

$$I'_\nu = I_\nu\{1 + 2fz\,\mathrm{Im}\,\chi^{(3)}\}.\qquad(3.90)$$

Separating the imaginary and real parts in $\chi^{(3)}$, that is, writing $\chi^{(3)} = \chi'^{(3)} + i\chi''^{(3)}$, we find that

$$I'_\nu = I_\nu(1 + 2f\chi''^{(3)}z).\qquad(3.91)$$

If we allow for the fact that the intensity of the Stokes wave is proportional to the number of Stokes photons, $N_\nu(z)$, we arrive at a similar expression for $N_\nu(z)$:

$$N_\nu(z) = N_\nu(0)\left(1 + 2f\chi''^{(3)}z\right). \tag{3.92}$$

If this formula is compared with the expansion of (3.76) for small z, which has the form

$$N_\nu(z) = N_\nu(0)(1 + G_\nu z), \tag{3.93}$$

we see that the gain can be expressed in terms of the imaginary part of the nonlinear susceptibility thus:

$$G_\nu = 2f\chi''^{(3)} = 4\pi N(\nu/c)\chi''^{(3)}|E_\omega|^2. \tag{3.94}$$

[where we have also allowed for (3.87)]. As can easily be seen, this formula is identical to the expression (3.72) obtained earlier, which is to be expected if we take into account the formula (3.77) that connects $\chi^{(3)}$ and $\chi^{(1)}_{np}$.

3.3.7. Exponential Law of Excitation for Stimulated Raman Scattering

The solution (3.85) is applicable for small distances z, when the variation δE_ν caused by the nonlinear susceptibility $\chi^{(3)}$ is small compared to E_ν.

Now let us turn to the case of large distances z. The unperturbed-field approximation becomes invalid for the field E_ν, but we will take it as still valid for E_ω, assuming that only a small fraction of the energy of the field E_ω is transferred at the given distances z to the excited Stokes wave of stimulated Raman scattering with frequency ν.

The equations of Section 3.3.6 can easily be modified to incorporate the large variation in E_ν. To this end, obviously, we must again use Maxwell's equation (3.84), but in the right-hand side of this equation we must assume E_ν to be a function of z that has yet to be found (rather than a given quantity). On the left-hand side of Eq. (3.84), instead of $\delta E_\nu(z)$ we will have the same field $E_\nu(z)$, since here we are seeking not the variation δE_ν but the field E_ν itself. Thus, instead of (3.84) we arrive at a linear Maxwell equation in the following form:

$$\frac{\partial^2 E_\nu(z)}{\partial z^2} + k_\nu^2 E_\nu(z) = -4\pi N\left(\frac{\nu}{c}\right)^2 \chi^{(3)}|E_\omega|^2 E_\nu(z). \tag{3.95}$$

This equation shows that a role played by the nonlinear susceptibility $\chi^{(3)}$

[in the right-hand side of (3.95)] is reduced to modifying the wave number k_ν of the field $\mathbf{E}_\nu(z)$:

$$k_\nu^2 \to k_\nu^2 + 4\pi N \left(\frac{\nu}{c}\right)^2 \chi^{(3)}|\mathbf{E}_\omega|^2. \tag{3.96}$$

Since the second term on the right-hand side is small compared to the first, by extracting the square root and assuming that $n_\nu \approx 1$ we find that

$$k_\nu \to k_\nu\left[1 + 2\pi N\chi^{(3)}|\mathbf{E}_\omega|^2\right], \tag{3.97}$$

which agrees with the general result (3.52). If we carry out the substitution (3.96) or (3.97), Maxwell's equation (3.95) assumes the form of a common wave equation. Its solution is a wave propagating along the z axis and having a wave number (3.97), that is,

$$\mathbf{E}_\nu(z) = \mathbf{E}_\nu(0)\exp\left(-ik_\nu' z + \frac{G_\nu z}{2}\right). \tag{3.98}$$

Here we have allowed for the formula (3.94) for the gain G_ν [which is equivalent to (3.72)] and have introduced the notation

$$k_\nu' = k_\nu\left[1 + 2\pi N\chi'^{(3)}|\mathbf{E}_\omega|^2\right]. \tag{3.99}$$

The fact that $\mathbf{E}_\nu(z)$ grows exponentially with z agrees (as it should) with the formula (3.76) for the number of photons, N_ν, as a function of z. As for (3.99), this result corresponds, obviously, to nonlinear variation in the refractive index of the medium,

$$n_\nu(E_\omega) = n_\nu + 2\pi N\chi'^{(3)}(\nu; \omega, \nu, -\omega)|\mathbf{E}_\omega|^2, \tag{3.100}$$

and is caused by the nonlinear susceptibility $\chi^{(3)}$. This agrees with the general formula for the refractive index

$$n_\nu = 1 + 2\pi N\chi_\nu, \tag{3.101}$$

where χ_ν is the exact atomic susceptibility at frequency ν [see also Eqs. (3.50) to (3.53)]. Replacing χ_ν and $\chi^{(1)}(\nu; \nu)$ in (3.101), we arrive at the linear refractive index (1.51). Equation (3.100) corresponds to the next term in the expansion of χ_ν in powers of the intensity of the incident wave, $I_\omega \propto |\mathbf{E}_\omega|^2$, an expansion that follows from the formula (2.1) for the polarization. The proportionality factor in such an expansion is simply

$\chi^{(3)}(\nu; \omega, \nu, -\omega)$. Finding the real part $\chi'^{(3)}$ from (3.77), we arrive at the final expression for (3.100):

$$n_\nu(E_\omega) = n_\nu + \frac{2\pi N |\chi_{np}^{(1)}(\nu; \omega)E_\omega|^2 (\nu - \omega + \omega_{pn})}{(\nu - \omega + \omega_{pn})^2 + \Gamma_p^2}. \qquad (3.102)$$

At exact resonance the nonlinear correction to the refractive index vanishes, while the gain G_ν defined in (3.72) attains its maximum and is equal to (3.73).

Of course, in the limiting case of small z, Eq. (3.98), as expected, transforms into (3.86), which corresponds to the linear mode of excitation.

3.3.8. Saturation of Stimulated Raman Scattering

The field $E_\nu(z)$ grows, according to (3.98), in an exponential manner as long as the unperturbed-field approximation for E_ω is valid, that is, as long as the energy transfer from field E_ω to field E_ν is low.

For large values of z, instead of one equation (3.95) we must solve a system of two coupled Maxwell equations: one for the field with frequency ν and the other for the field with frequency ω. The second is written in a manner similar to Eq. (3.95). An example of an analytical solution of such a system is given in Loudon, 1973.

We will not formulate and solve the system of the coupled equations, however. Rather, we will consider only the limit where z is so large that all dependence on z vanishes. Such a limit is known as the *saturation mode*. The characteristics of the radiation at ν in this mode can be obtained without resorting to the Maxwell equations. Instead, the simpler law of energy conservation is employed.

All the Feynman diagrams used up to this point show that in any process of stimulated Raman scattering or spontaneous Raman scattering, the total number of photons in the medium is conserved. For instance in spontaneous Raman scattering, one photon of the field with frequency ω disappears, and one photon of the field with frequency ν is created. The number of photons in the incident wave of frequency ω is proportional to the ratio I_ω/ω, where I_ω is the intensity of this wave. Thus, the law of conservation of the total number of photons can be written in a form valid for all values of z:

$$\frac{I_\omega(z)}{\omega} + \frac{I_\nu(z)}{\nu} = \frac{I_\omega(0)}{\omega}. \qquad (3.103)$$

In the saturation mode, when $z \to \infty$, we have $I_\omega(\infty) = 0$, which corresponds to the complete transfer of energy from the field with frequency ω. Hence, we find that

$$I_\nu(z) = \frac{\nu}{\omega} I_\omega(0) < I_\omega(0). \qquad (3.104)$$

The remaining part of the intensity of the wave of frequency ω, equal to $(1 - \nu/\omega)I_\omega(0)$, is absorbed by the medium. Such absorption is caused by the excitation of the atoms of the medium into a real populated state p. Obviously, there are relations similar to (3.104) for the power and total energy of radiation of frequency ν.

Note that the rate with which the saturation mode is achieved depends on the width Γ_p of state p, which is determined by spontaneous cascade relaxation into initial state n. An increase in the relaxation time obviously leads to a slowing down of the transition to saturation.

The dependence of stimulated Raman scattering on z is difficult to measure in experiments. However, the same dependence can be observed simply by using a gas cell of fixed length and increasing the pressure of the gas, that is, increasing the number of atoms per unit volume, N. The results of such an experiment should give a correct picture of how $E_\nu(z)$ varies with N, since G_ν given by (3.72) is proportional to N.

An example of a study of the process of stimulated Raman scattering is the work of Bloembergen, Breit, Lallemand, Pine, and Simova, 1967, in which the yield of stimulated Raman scattering was studied as a function of the pressure of molecular hydrogen in a gas cell. Figure 3.18 demonstrates that the dependence agrees qualitatively with that discussed above, namely, exponential growth with N and saturation at large values of N. The results of the paper mentioned imply that the experimental data are in good agreement with the theoretical results.

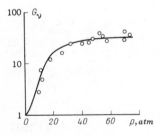

Figure 3.18. The gain of stimulated Raman scattering as a function of gas pressure (Bloembergen, Breit, Lallemand, Pine, and Simova, 1967). The solid curve corresponds to theoretical results; the plotted points represent experimental data.

3.3.9. One-Photon Absorption as a Competing Process

We have already discussed the merits of resonant stimulated Raman scattering, that is, the case where a one-photon resonance appears at the frequency ω of the incident wave between the initial state n and an intermediate state m. In the event of a resonance the gain increases, and its maximum value is given by (3.75).

This, however, is accompanied by a resonant increase in the probability of a competing process, namely, one-photon absorption of the wave with frequency ω that transfers the atom into the excited state m. This process has been studied in detail in Section 1.4, as well as in Section 3.2.6, where it was considered as competitive with the excitation of the third harmonic. Obviously, it dominates over stimulated Raman scattering if the time Γ_m^{-1} of one-photon or cascade spontaneous deexcitation of state m accompanied by transition to the ground state n is short compared to the time of stimulated emission of a Stokes photon in stimulated Raman scattering with frequency ν.

To arrive at numerical estimates we recall (see Section 1.4 for details) that, due to one-photon absorption, the intensity and hence the number of photons in the incident radiation decrease (according to Bouguer's law) exponentially with z, which is the distance traveled by the incident wave of frequency ω [see Eq. (1.68)]:

$$N_\omega(z) = N_\omega(0)\exp(-\kappa_\omega z). \tag{3.105}$$

According to (1.69) and (1.25) the one-photon absorption coefficient κ_ω is given by the following formula:

$$
\begin{aligned}
\kappa_\omega &= -4\pi N k_\omega \mathrm{Im}\, \chi^{(1)}(\nu; \omega) \\
&= 4\pi N k_\omega (\mathbf{r}_{mn} \cdot \mathbf{e})^2 \Gamma_m \left[(\omega_{mn} - \omega)^2 + \Gamma_m^2 \right]^{-1},
\end{aligned} \tag{3.106}
$$

where $\mathbf{e}$ is the unit polarization vector of the incident wave.

Our goal is to take into account one-photon absorption in the process in which the intensity of the Stokes wave in stimulated Raman scattering is increased, that is, to modify (3.76) for $N_\nu(z)$ in such a manner that it incorporates κ_ω.

Equation (3.105) remains valid as long as the energy transfer from the incident wave of frequency ω to the Stokes wave in stimulated Raman scattering is small, so that the decrease in the number of photons of frequency ω is determined primarily by one-photon absorption. According to (3.104) this is always the case if the Stokes frequency ν is sufficiently low compared to the frequency ω of the incident wave. Substituting (3.105) into (3.69), we find that G_ν becomes a decreasing function of z:

$$G_\nu(z) = G_\nu(0)\exp(-\kappa_\omega z).\qquad(3.107)$$

Here $G_\nu(0)$ is determined by (3.72). Now, if we substitute (3.107) into (3.69), we arrive at an equation for the number of photons in the Stokes wave:

$$dN_\nu = G_\nu\exp(-\kappa_\omega z)N_\nu(z)\,dz.\qquad(3.108)$$

This equation has a simple solution:

$$N_\nu(z) = N_\nu(0)\exp\left\{\frac{G_\nu}{\kappa_\omega}[1 - \exp(-\kappa_\omega z)]\right\}.\qquad(3.109)$$

For $\kappa_\omega z \ll 1$ this yields (3.76), as expected in the absence of one-photon absorption. In the opposite limiting case, $\kappa_\omega z \gg 1$, the solution (3.109) yields (for a long cell containing the gas, namely, one whose length L is much greater than κ_ω^{-1})

$$N_\nu(\infty) = N_\nu(0)\exp\left(\frac{G_\nu}{\kappa_\omega}\right).\qquad(3.110)$$

This formula clearly demonstrates that the intensity of a stimulated-Raman-scattering wave with frequency ν increases, that is, $N_\nu(\infty) > N_\nu(0)$, if

$$G_\nu > \kappa_\omega.\qquad(3.111)$$

If we replace κ_ω with its expression (3.106) and G_ν with (3.74), and cancel out identical factors, we arrive at the following inequality:

$$\nu|\mathbf{r}_{pm} \cdot \mathbf{r}_{mn}||\mathbf{E}_\omega|^2 > \omega\Gamma_p\Gamma_m.\qquad(3.112)$$

This condition imposes on the field strength $\mathbf{E}_\omega$ in the incident radiation a fairly high threshold, below which one-photon absorption is dominant over stimulated Raman scattering. On the other hand, this condition places an upper bound on the value of Γ_m and means, essentially, that the rate at which state m is deexcited after one-photon absorption is low (see the beginning of Section 3.3.9). Correspondingly, the quantity

$$\tau_{\text{st.R.s.}} = \frac{\omega}{\nu}\frac{\Gamma_p}{\mathbf{r}_{pm} \cdot \mathbf{r}_{mn}}|\mathbf{E}_\omega|^{-2}\qquad(3.113)$$

can be called the time of stimulated (induced) emission of a Stokes photon in stimulated Raman scattering with frequency ν.

The condition (3.112) remains valid qualitatively even if the distribution function of the absorbed photons, $g(\omega)$, is not Lorentzian, which would be the case in collision broadening [this fact was used in the formula (3.106) for κ_ω], but Gaussian, specifically for Doppler broadening. Then for Γ_m we must simply take the Doppler linewidth. A typical situation is the presence of Doppler broadening at the center of a line and collision broadening in the wings of the line.

Although quantitatively the solution (3.109) becomes invalid at $\nu \sim \omega$, when the energy transfer from the wave with frequency ω to the Stokes wave with frequency ν is considerable, qualitatively the condition (3.112) for the stimulated-Raman-scattering threshold remains valid. We see that this condition can be realized also be selecting a transition with high values of the dipole matrix elements $\mathbf{r}_{pm}$ and $\mathbf{r}_{mn}$.

The condition (3.112) was obtained on the assumption that the cell with the gaseous medium is long. If we denote the length of the cell by L and ignore one-photon absorption, then, according to (3.76), the maximum increase in the intensity of the Stokes wave in stimulated Raman scattering is $\alpha = \exp(G_\nu L)$. If we define the threshold of $\mathbf{E}_\omega$ as the value corresponding to a certain α (say, $\alpha = 10$), we see that this threshold is determined in accordance with the formula (3.74) for G_ν. The true threshold of $\mathbf{E}_\omega$ is the maximum of these values and is determined from the estimate (3.112).

3.3.10. Excitation of Stimulated Raman Scattering by Nonmonochromatic Radiation

Up till now we have assumed that the incident wave of frequency ω is monochromatic. Deviations from monochromaticity, obviously, are important only when there are one or more resonances in the excitation of a Stokes wave in stimulated Raman scattering. Here we are speaking of the one-photon resonance at $\omega \approx \omega_{mn}$, which ensures that there is resonant stimulated Raman scattering. Hence, we assume that $|\omega_{mn} - \omega| \ll \omega$.

As for the mechanism of this nonmonochromaticity, we will assume, without going into details, that the spectral width of the incident radiation is $\Delta\omega$. Another assumption is that $\Delta\omega \gg \Gamma_m$, with Γ_m the width of the resonant excited state m. This means that the characteristic response time of the medium (of the order of Γ_m^{-1}) is large compared to the characteristic time of fluctuations of the field in the incident wave (of the order of $\Delta\omega^{-1}$). Under these conditions the medium responds to the average intensity of the wave, $\langle I_\omega \rangle$, and hence, in the expression (3.74) for G_ν we must replace $|\mathbf{E}_\omega|^2$ with $\langle |\mathbf{E}_\omega| \rangle^2$, which is proportional to $\langle I_\omega \rangle$.

Now let us discuss the role of the resonance denominator $(\omega_{mn} - \omega)^2 + \Gamma_m^2$ in the expression (3.74) for G_ν. As noted in Section 1.2.4, for the width in resonance denominators we must substitute the maximum width, which in the case at hand is $\Delta\omega$. With this in mind, we can write the resonance denominator in (3.74) as $(\omega_{mn} - \omega)^2 + (\Delta\omega)^2$ (to within an order of magnitude), and if we allow for (3.36), we have

$$G = 8\pi^2 N \Gamma_p \left(\frac{\nu}{c^2} \right) \frac{\left| (\mathbf{r}_{pm} \cdot \mathbf{e})(\mathbf{r}_{mn} \cdot \mathbf{e}_0) \right|^2 \langle I_\omega \rangle}{\left(\omega_{mn} - \omega \right)^2 + \left(\Delta\omega \right)^2}. \qquad (3.114)$$

A conclusion that can be drawn from this is that for small misfits (detunings) $|\omega_{mn} - \omega| \ll \Delta\omega$ in broad-band radiation ($\Delta\omega \gg \Gamma_m$) the gain G_ν is independent of the detuning. A given value of G_ν (say, the threshold value) is reached for an average incident intensity $\langle I_\omega \rangle$ independent of the misfit, $\omega_{mn} - \omega$.

If the misfits are great, that is, $|\omega_{mn} - \omega| \gg \Delta\omega$, we can neglect $\Delta\omega$ in the resonance denominator of (3.114), with the result that a given value of G_ν (say, the same threshold value as the one mentioned above) is reached at an average incident intensity that grows with the misfit according to the law $\langle I_\omega \rangle \propto (\omega_{mn} - \omega)^2$.

As an example illustrating these theoretical considerations we cite the experimental study conducted by Korolev, Mikhailov, and Odintsov, 1978, who investigated stimulated Raman scattering in rubidium vapor for various $\Delta\omega$. The excited state $5P$ was taken as the initial state n. The incident radiation was tuned in resonance to the higher excited state $m = 5D$. After emitting a Stokes photon in the course of stimulated Raman scattering, the atom went over to the excited state $p = 6P$.

Figure 3.19 shows the threshold incident intensity $\langle I_\omega \rangle$ as a function of $\omega_{mn} - \omega$ for different values of $\Delta\omega$. For a small spectral width, say $\Delta\omega = 0.2$ cm^{-1}, the intensity obeys the quadratic law $\langle I_\omega \rangle \propto (\omega_{mn} - \omega)^2$ predicted by the theory. In this case the incident wave can be considered

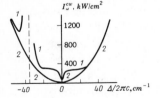

Figure 3.19. The threshold incident intensity for stimulated Raman scattering as a function of the misfit (detuning) $\Delta = \omega_{mn} - \omega$ for different linewidths $\Delta\omega$ of the incident radiation: curve 1, $\Delta\omega = 20$ cm^{-1}; curve 2, $\Delta\omega = 0.2$ cm^{-1}.

monochromatic. On the other hand, for a large spectral width, say $\Delta\omega =$ 20 cm^{-1}, in the interval of detunings $|\omega_{mn} - \omega| < 20$ cm^{-1}, the value of the threshold intensity $\langle I_\omega \rangle$ is independent of $\omega_{mn} - \omega$ (the plateau in Figure 3.19). At larger values of $|\omega_{mn} - \omega|$ the quadratic dependence again takes over.

In the immediate vicinity of the resonance at $\omega = \omega_{mn}$, as Figure 3.19 shows, there is a considerable drop in the threshold intensity $\langle I_\omega \rangle$ for a broad spectrum ($\Delta\omega = 20$ cm^{-1}). The resonant drop in the threshold is caused by the fact that only a small fraction of the frequency components of the incident-radiation spectrum prove to be in exact resonance. This is equivalent to an effective decrease of $\Delta\omega$ (the radiation becomes more monochromatic), which leads to a drop in the threshold intensity $\langle I_\omega \rangle \propto (\Delta\omega)^2$.

As for the resonant increase in the threshold intensity $\langle I_\omega \rangle$ at $\omega_{mn} - \omega$ $= -35$ cm^{-1}, which can be seen in Figure 3.19, it is caused by the fact that at the given frequency there is a two-photon resonance $2\omega = \omega_{mq}$, where $q = 5S$ is the ground state of the rubidium atom. As a result, two-photon absorption (Section 3.7) is dominant over the stimulated-Raman-scattering threshold.

A detailed analysis of stimulated Raman scattering in a randomly varying field of the incident wave has been carried out by Akhmanov, 1974.

3.3.11. Stimulated Hyper-Raman Scattering

As noted in Section 1.1, in spontaneous Raman scattering one photon of the incident wave with frequency ω is absorbed, but obviously a process is possible in which the atom absorbs two photons of the incident wave, after which it emits a spontaneous photon of frequency ν and goes over to its final state, which in general may differ from the initial state. This is known as *spontaneous hyper-Raman scattering*. The absorption and emission of photons in this process is shown in Figure 3.20, while the Feynman diagram for the corresponding susceptibility $\chi_{np}^{(2)}(\nu; \omega, \omega)$ is shown in Figure 3.21. According to the law of conservation of energy, we have the following relationship for the frequency of the spontaneously emitted photon: $\nu = 2\omega - \omega_{pn}$. This, in contrast to spontaneous Raman scattering, signifies that this photon is not necessarily a Stokes photon ($\nu < \omega$; Figure 3.20a), that is, it can be an anti-Stokes photon ($\nu > \omega$; Figure 3.20b) even when the initial state n is the ground state.

Note that the frequencies of the two photons absorbed by an atom can differ ($\omega_1 \neq \omega_2$), and the same is true of the field strengths in the corresponding electromagnetic waves. In this case one must deal with the nonlinear susceptibility $\chi_{np}^{(2)}(\nu; \omega_1, \omega_2)$. The tensor structure of this quantity

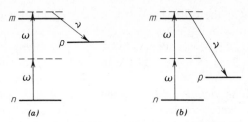

Figure 3.20. Emission-absorption diagrams for spontaneous hyper-Raman scattering: (*a*) emission of a Stokes photon, and (*b*) emission of an anti-Stokes photon.

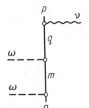

Figure 3.21. The Feynman diagram for the amplitude of spontaneous hyper-Raman scattering.

has been considered by Manakov and Ovsyannikov, 1980, and we will not dwell on it here.

When the intensity of the incident wave with frequency ω (or waves with frequencies ω_1 and ω_2) is high, the emission of a photon with frequency ν becomes stimulated, as in the case when spontaneous Raman scattering is replaced by stimulated Raman scattering. Such a process is known as *stimulated hyper-Raman scattering.* This process is determined by the imaginary part of the fifth-order susceptibility $\chi^{(5)}(\nu; \omega, \omega, \nu, -\omega, -\omega)$, while stimulated Raman scattering is determined, the reader will recall, by the imaginary part of the cubic susceptibility $\chi^{(3)}(\nu; \omega, \nu, -\omega)$. One of the Feynman diagrams representing the process is depicted in Figure 3.22.

Both spontaneous hyper-Raman scattering and stimulated hyper-Raman scattering, since they require the absorption of two photons from the incident radiation, are processes of the next order of perturbation theory beyond spontaneous Raman scattering and stimulated Raman scattering. This, quite obviously, requires a higher field strength in the incident wave.

Figure 3.22 shows that $\chi^{(5)}$, and hence the imaginary part of $\chi^{(5)}$ and the gain G_ν for stimulated hyper-Raman scattering, are proportional to the square of the incident intensity I_ω (the reader will recall that $G_\nu \propto I_\omega$ in stimulated Raman scattering). In the diagram in Figure 3.22 there is no summation over state p, since $\omega_{pn} \approx 2\omega - \nu$ because of the resonance condition. The value of Im $\chi^{(5)}$ and hence that of G_ν are proportional to $\Gamma_p[(2\omega - \nu - \omega_{pn})^2 + \Gamma_p^2]^{-1}$.

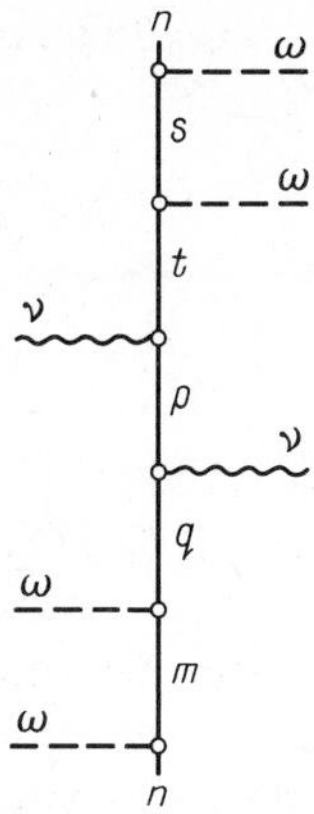

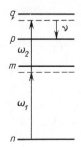

Figure 3.22. A Feynman diagram for the nonlinear susceptibility in stimulated hyper-Raman scattering.

Note that in stimulated hyper-Raman scattering (just as in stimulated Raman scattering) the phase-matching condition is satisfied identically, since according to Figure 3.22 this condition has the form $2\mathbf{k}_\omega + \mathbf{k}_\nu = 2\mathbf{k}_\omega + \mathbf{k}_\nu$. The same is true for the case of two different incident waves with frequencies ω_1 and ω_2.

As an example we will describe the experiment in stimulated hyper-Raman scattering conducted by Reif and Walther, 1978. Strontium vapor (the ground state with two valence electrons, $n = 5S^2$) was irradiated by the light of two dye lasers. As a result the two-photon transition into the one-electron state $q = 5S5D$ was resonantly excited ($\omega_{qn} \approx \omega_1 + \omega_2$), after which the atom emitted a Stokes photon in stimulated hyper-Raman scattering with a wavelength $\lambda = 16\ \mu$m and went over to the one-electron state $p = 5S6P$ (Figure 3.23). The detuning from resonance was also relatively small when one photon from the incident wave of frequency ω_1 was absorbed, which resulted in the excitation of the one-electron state $5S5P$. Thus, in each stage of the atomic transitions either resonance or quasiresonance conditions were realized. This means that in the expression for $\chi^{(5)}$ (see the Feynman diagram in Figure 3.22) there is no summation

Figure 3.23. Atomic transitions in an atom of strontium in the excitation of stimulated hyper-Raman scattering (Reif and Walther, 1978).

over the intermediate states m, q, p, t, and s, with $t = q$ and $s = m$. Thus, $\chi^{(5)}$ assumes the simple form

$$\chi^{(5)}(\nu; \omega, \omega, \nu, -\omega, -\omega)$$

$$= \frac{[z(5S - 5P)z(5P - 5D)z(5D - 6P)]^2}{[\{\omega(5P - 5S) - \omega\}\{\omega(5D - 5S) - 2\omega\}]^2}$$

$$\times [\omega(6P - 5S) - 2\omega + \nu + i\Gamma(6P)]^{-1}, \qquad (3.115)$$

where in parentheses we give the states of only one valence electron undergoing the transition. The measured resonant dependence of the gain of the Stokes radiation, $G_\nu \propto N\chi''^{(5)}I_\omega^2$, on ν agrees qualitatively with (3.115).

Obviously, in principle there may be processes of higher order in the number of incident photons absorbed by the atom (3ω, 4ω, etc.), but the probability of such processes drops drastically with increasing order.

3.3.12. Stimulated Raman Scattering from Excited Atomic States

Up till now we have assumed that in the majority of cases the atom under consideration was in the ground state at the initial moment. Stimulated Raman scattering, however, may take place for excited atoms or molecules, too. To observe this, of course, the initial excited state n must be highly populated, a situation made possible thanks to the method of optical pumping (see Bertein, 1969). As in the previous case, resonant stimulated Raman scattering is of greatest interest. In such a resonance process the frequency of the incident wave is close to the frequency of an atomic transition. In what follows we will discuss only one-photon resonance.

The obvious quantitative difference between stimulated Raman scattering from excited states and from the ground state is the greater linewidth of the former radiation due to the presence of a width in the initial excited state n. But there are also two qualitative differences. The first belongs to resonant stimulated Raman scattering and consists of the emergence of a new spectral line (see the similar case for spontaneous Raman scattering in Section 1.2). We start with the common scheme of stimulated Raman scattering. An atom in the initial excited state n absorbs a single photon from the incident wave of frequency ω and goes over to a state m, after which it spontaneously emits a photon of frequency ν and goes over to the final state p. Such spontaneous Raman scattering is the initial stage in the process of excitation of stimulated Raman scattering at the frequency ν of the spontaneous transition. However, since state n possesses a finite width,

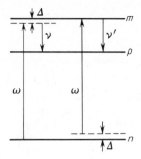

Figure 3.24. Resonant stimulated Raman scattering (frequency ν) and hot stimulated radiation (frequency ν') for stimulated Raman scattering from excited states of atoms.

the absorption of a photon from the initial wave of frequency ω may occur not only from the center of the spectral line of state n, as was just considered (the left side of Figure 3.24) but also from a quasi energy level lying higher or lower than level n by a small quantity $\Delta = \omega_{mn} - \omega$ if the broadening lies within this interval (the right side of Figure 3.24). The decrease of the population of the initial state n, which occurs when we move away from the center of the spectral line, is compensated for by the fact that the resonance condition is satisfied with a greater accuracy by tuning to the intermediate level m. As a result of photon of frequency $\nu' = \omega_{mp}$ is spontaneously emitted in the process. This emission gives a new spectra line.

The competitiveness of stimulated Raman scattering at frequency ν' is considerably higher when the width of state n is great, namely, when it is comparable to or greater than Δ. Only then are the populations of the center of the line of state n and of its quasi energy level comparable. At the same time the width of level m must be small compared to Δ, so as to lower the probability of exciting state m according to the scheme in the left half of Figure 3.24, which corresponds to stimulated Raman scattering at ν. Thus, for $\Gamma_n \gg \Gamma_m$, stimulated Raman scattering is excited primarily at frequency ν', while for $\Gamma_n \ll \Gamma_m$ it occurs at ν (for a similar result in the case of spontaneous Raman scattering see Section 1.2.5). The latter case includes stimulated Raman scattering from the ground state which we have discussed earlier. The emission of radiation of frequency ν is commonly known as *resonant stimulated Raman scattering*, and the radiation of frequency ν' is known as *hot stimulated radiation* (Shen, 1977). Obviously, the two frequencies coincide at $\Delta = 0$, that is, to put it precisely, they cannot be separated when the misfit becomes smaller than Γ_n and Γ_m.

Similar results hold for resonant stimulated hyper-Raman scattering when $\omega_1 + \omega_2 \approx \omega_{qn}$. Here ω_1 and ω_2 are the frequencies of the incident waves. But if stimulated hyper-Raman scattering is a resonance process where one photon is absorbed, that is, $\omega_1 \approx \omega_{mn}$, then we are dealing with

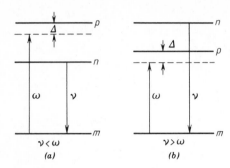

Figure 3.25. Other emission-absorption diagrams for stimulated Raman scattering from an excited atomic state, n, accompanied by (a) Stokes and (b) anti-Stokes photons.

three atomic levels n, m, and q placed in two strong resonance fields of frequencies ω_1 and ω_2. In this case around every state there appear three quasi energy levels (Kraĭnov, 1977); the number of spectral lines of stimulated hyper-Raman scattering in the Stokes transition $q \to p$ increases, too. The intensities of these lines have been calculated by Mavroyannis, 1983.

The second qualitative feature of stimulated Raman scattering from excited states that sets it apart from stimulated Raman scattering from the ground state is the possibility of realizing the case where the intermediate state m lies below the initial excited state n (Figure 3.25). The spontaneous transition from n to m is then accompanied by resonant excitation from state m into state p produced by the incident wave of frequency ω. We see that in view of the low population of the intermediate state m there is no competition from one-photon absorption in the $m \to p$ transition (Bobovich and Bortkevich, 1971). The scheme in Figure 3.25a corresponds to the excitation of a Stokes photon in stimulated Raman scattering ($\nu < \omega$), although it is clear that there may be anti-Stokes photons excited in stimulated Raman scattering ($\nu > \omega$), which is clearly impossible for stimulated Raman scattering from the ground state. The anti-Stokes case is depicted in Figure 3.25b.

3.3.13. Stimulated Raman Scattering in Other Media

There is no doubt that stimulated Raman scattering can occur in molecular gases, plasmas, liquids, and solids, in addition to atomic gases. The main features of these processes are the same as for atomic gases. The chief difference here lies in the fact that other degrees of freedom are excited and the scattered-radiation frequencies lie in other ranges. We will briefly discuss the main features of stimulated Raman scattering in such media (see Martin and Falicov, 1975).

Let us start with molecular gases. In principle, stimulated Raman scattering may manifest itself on both vibrational and rotational transitions.

Both cases have been observed, for instance, in H_2 (Bloembergen, Breit, Lallemand, Pine, and Simova, 1967). However, the main characteristics of stimulated Raman scattering on vibrational transitions differ from those on rotational transitions. This is true for the thresholds of excitation of stimulated Raman scattering and for the dependence of the threshold on the type of molecule and the pressure of the gas. Some data on stimulated Raman scattering in molecular gases have been summed up by Averbakh, Betin, Gaponov, Makarov, and Pasmanik, 1978, who also give references to original papers.

Stimulated Raman scattering has been repeatedly observed in various liquids. We must first note liquefied gases, in particular, liquid nitrogen and oxygen, which are being successfully used in amplifiers and generators based on stimulated Raman scattering (see below). Moreover, stimulated Raman scattering has been observed in benzene, nitrobenzene, carbon disulfide, and many other liquids. It must be noted that stimulated-Raman-scattering studies are usually hindered by the self-focusing of the exciting radiation (Section 3.5). Self-focusing changes the spatial distribution of the radiation incident on the liquid and thereby changes the radiation intensity, which leads to a change in the threshold of stimulated Raman scattering and other characteristics of the scattering process. Some details of stimulated Raman scattering in liquids have been discussed by Kaiser and Maier, 1972.

In solids (transparent insulators) stimulated Raman scattering occurs via different mechanisms. One of these is the interaction of the exciting radiation with the phonons in the solid. When this radiation interacts with lattice vibrations in polar ionic crystals, stimulated Raman scattering manifests itself in the excitation of polaritons. The main principles of stimulated Raman scattering on polaritons has been discussed by Shen, 1975a. Another mechanism consists in the excitation of electronic transitions in solids. These are transitions of conduction electrons between Zeeman sublevels accompanied by spin flip, transitions that occur in some semiconductor crystals. The spin-flip frequency, which determines the change in frequency due to scattering, is proportional to the strength of the field in the incident wave. This enables varying the frequency of stimulated Raman scattering on spin sublevels (Shen, 1975a).

In plasmas, stimulated Raman scattering appears during the excitation of plasma waves of various kinds (Phillion, Banner, Campbell, Turner, and Estabrook, 1982).

It must be noted that on the whole the situation with condensed and dense media is much more complicated than with rarefied gases. In dense media other types of stimulated scattering, together with stimulated Raman scattering, play an important role. These are Brillouin scattering on the fluctuations of the density of the medium (Fabelinskiĭ, 1968; Starunov and Fabelinskiĭ, 1970; Zel'dovich and Sobelman, 1970), stimulated temperature (entropy) scattering on the fluctuations of the entropy of the medium (Starunov and Fabelinskiĭ, 1970; Zel'dovich and Sobelman, 1970), stimulated concentration scattering on concentration fluctuations (Cummings, Knable, and Yeh, 1964), and stimulated scattering in the Rayleigh-line wing on fluctuations of the anisotropy of molecules (Mash, Morozov, Starunov, and Fabelinskiĭ, 1965). The ratio between the thresholds for the different types of stimulated scattering is determined by the specific properties of the medium. The competition between various types of scattering makes it difficult to isolate stimulated Raman scattering or even to observe it.

3.3.14. Stimulated Raman Scattering in Nonlinear Optics and Quantum Radiophysics

Stimulated Raman scattering is one of the main mechanisms through which the energy of the incident radiation is dissipated in the interaction with transparent media. However, there are other interesting aspects of stimulated Raman scattering, aspects relating it to applications in nonlinear optics and quantum radiophysics. The first is the possibility of using stimulated Raman scattering to generate high-power coherent infrared radiation with alternating wavelength and to carry out spectroscopic studies. The merits of stimulated Raman scattering lie primarily in the fact that no phase matching is required to realize this process.

Let us start with sources of infrared radiation. At the base of this application lies stimulated Raman scattering in vapors of elements that have low-lying excited levels, say, vapors of alkali and alkaline-earth metals. To excite stimulated Raman scattering, the radiation of dye lasers in the ultraviolet frequency range is employed. For instance, in potassium, stimulated Raman scattering has been excited according to the following scheme: radiation of a dye laser excites a potassium atom from the ground state $4S$ into state $5P$, which relaxes into state $5S$. The $5P \to 5S$ transition, on which stimulated Raman scattering is excited in this case, corresponds to a wavelength of approximately 2.7 μm. Varying the frequency of the exciting radiation and hence the misfit on the $4S \to 5P$ transition enables varying the frequency of stimulated Raman scattering. It was found that the latter frequency could be varied within 200 cm^{-1}, which corresponds to wavelengths ranging from 2.6 to 2.8 μm. Such schemes have been realized using many atoms, so that it has been possible to construct sources of coherent infrared radiation with wavelengths ranging from several micrometers to several tens of micrometers, with the possibility of varying the frequency up to several thousand cm^{-1}. Since the transformation coefficient of the incident radiation in stimulated Raman scattering may reach a value of the order of unity, the intensity of such sources of infrared radiation is determined primarily by the intensity of the exciting radiation. The practical value of such sources is obvious, especially for the infrared spectroscopy of molecules.

Building a laser with a lasing frequency corresponding to the stimulated-Raman-scattering frequency is for a number of reasons more promising than exciting stimulated Raman scattering in the usual way. In a laser one is able to attain a smaller beam divergence, a greater brightness at stimulated-Raman-scattering frequency, and a higher critical intensity of the first Stokes component, an intensity at which the energy transfer to this component is still insignificant. To construct a stimulated-Raman-scattering laser, the active medium is placed in a Fabry-Perot resonator with mirrors whose reflectivities are maximal for the first component and minimal for the second. This can be achieved because the frequencies of the two components differ by a considerable quantity, lying, as a rule, in between 10^2 and 10^{-3} cm^{-1}. Thus, the ratio between the intensities of the first and second components can be made much higher by lasing than by exciting stimulated Raman scattering in the same medium. The divergence of the radiation produced by a Raman laser is determined, as for any laser, by the design of the laser cavity, and is independent of the geometry of the exciting radiation. Therefore, even with multimode exciting radiation one can attain diffraction-limited divergence. Raman lasers have been constructed with many active media: compressed gases, liquids, and transparent insulators; the wavelength of the emitted radiation lies in the range of 0.5 to 1.0 μm, while the spectral width is less than 1 cm^{-1} (Grasyuk, 1974).

Stimulated Raman scattering is widely used in nonlinear spectroscopy, too (see Akhmanov and Koroteev, 1981). However, although the intensity of stimulated Raman scattering is far greater than that of spontaneous Raman scattering, the direct use of stimulated Raman scattering in spectroscopy is complicated both by the excitation of many combination frequencies and by other nonlinear phenomena that appear in the medium under the action of the exciting radiation. For this reason direct use of stimulated Raman scattering in spectroscopy is reduced to amplifying spontaneous Raman scattering below the threshold of excitation of stimulated Raman scattering. This method is known as combination amplification (Akhmanov and Koroteev, 1981).

Another avenue in the application of stimulated Raman scattering in spectroscopy has been opened by the possibility of using ultrashort pulses of the radiation of picosecond lasers for exciting stimulated Raman scattering. The characteristic relaxation times for molecules are usually much longer than the duration of these pulses, so that picosecond excitation may be considered instantaneous. Auxiliary radiation is employed to observe the decay of molecular vibrations. Use of the time-dependent theory of stimulated Raman scattering (Akhmanov, Drabovich, Sukhorukov, and Chirkin, 1970) makes it possible to determine the relaxation times of molecules from the results of such experiments (Akhmanov and Koroteev, 1981).

The phenomenon of stimulated Raman scattering also lies at the base of a number of widely used methods of active spectroscopy (Akhmanov and Koroteev, 1981), including coherence anti-Stokes spectroscopy (Akhmanov and Koroteev, 1981). Strictly speaking, however, the four-photon interaction process is fundamental to these methods (Section 3.4), in view of which some information about them will be given in Section 3.4.

3.3.15. Conclusion

The above analysis of the main laws governing stimulated Raman scattering of light clearly shows the important features of this process, the differences that exist between it and the process of harmonic excitation, and the general aspects common to the two. In short, we must emphasize that stimulated Raman scattering has a spontaneous analog, so to say. The specific feature of the first components in stimulated Raman scattering (Stokes and anti-Stokes) as a nonlinear process is the stimulated nature of the relaxation of the excited atomic state. The linear susceptibility of the medium is responsible for the appearance of these components. The coherence of the process of excitation of the first components in stimulated Raman scattering and the coherence of the radiation at the respective frequencies are a consequence only of the coherence of the incident radiation. There is no requirement of phase matching in this case.

A more complicated situation arises when two waves propagate in the medium: the external exciting wave and a wave with the frequency of the first component of stimulated Raman scattering. Nonlinear susceptibility of the medium causes these two waves to interact. In this case, just as in any other case where waves with distinct frequencies interact in a medium, phase matching is required for energy to be transferred to the polarization wave. At this stage there is a qualitative analogy between this process and higher-harmonic excitation. The excitations of higher harmonics in stimulated Raman scattering are of a similar nature.

As for hyper-Raman scattering, although this process does have a spontaneous analog, both involve the absorption of two photons, that is, are essentially nonlinear.

In conclusion we note that the interaction of waves with distinct frequencies, a situation that accompanies the formation of higher harmonics in stimulated Raman scattering, is only a particular case of the general phenomenon of wave coupling in a nonlinear medium. We will devote the next section to this phenomenon.

3.4. COUPLED WAVES

3.4.1. Introduction

Let us take the simplest case of two waves with distinct frequencies ω_1 and ω_2 impinging on a nonlinear medium. This situation, it must be noted, is quite realistic. There is no need to employ two lasers for its realization. As shown in Section 3.3, when one wave of frequency ω_1 impinges on a nonlinear medium, stimulated Raman scattering is excited in the medium and there appears a second wave whose frequency is, say, ω_2. Thus, in the nonlinear medium there are two waves propagating in the same direction [the vectors $\mathbf{k}(\omega_1)$ and $\mathbf{k}(\omega_2)$ are collinear] and having frequencies ω_1 and ω_2.

Each of these two waves excites in the medium (independently of the other) polarization waves with frequencies $K_1\omega_1$ and $K_2\omega_2$, which are higher-harmonic frequencies. At the same time, however, there are polarization waves excited by the two waves simultaneously. The frequencies of these waves are equal to linear combinations of ω_1 and ω_2, that is, $\omega_i = K_1\omega_1 \pm K_2\omega_2$. Thus, the two incident waves with frequencies ω_1 and ω_2 become coupled and form a third wave with a frequency ω_i. For the sake of generality, all three waves are called *coupled*.

3.4.2. Wave Coupling in a Medium with Quadratic Nonlinearity

Let us consider the simplest case of two waves propagating in the half space $z < 0$ and impinging on a nonlinear medium that occupies the half space $z > 0$. We denote the electric field strengths in these waves by $\mathbf{E}_1 = \mathbf{E}(\omega_1)\exp(i\omega_1 t - i\mathbf{k}_1 \cdot \mathbf{r})$ and $\mathbf{E}_2 = \mathbf{E}(\omega_2)\exp(i\omega_2 t - i\mathbf{k}_2 \cdot \mathbf{r})$. The wave vectors $\mathbf{k}_1$ and $\mathbf{k}_2$ of these waves are not necessarily collinear and may differ in length. The wave numbers of these waves are

$$k_1 = \frac{n(\omega_1)\omega_1}{c}, \qquad k_2 = \frac{n(\omega_2)\omega_2}{c}. \tag{3.116}$$

Let us discuss the process of excitation by these two waves of electromagnetic waves in a medium with quadratic polarization. The polarization contains terms that are mixed in the vectors $\mathbf{E}_1$ and $\mathbf{E}_2$. Let us examine the intensities of these waves.

We start with the quadratic polarization of the form

$$P_i^{(2)} = \sum_{jk} \chi_{ijk}^{(2)} E_j(\omega_1) E_k(\omega_2) \exp[i(\omega_1 \pm \omega_2)t - i(\mathbf{k}_1 \pm \mathbf{k}_2) \cdot \mathbf{r}]. \quad (3.117)$$

From the results of Section 3.1 it follows that this polarization leads to the excitation of two electromagnetic waves with frequencies $\omega_\pm = \omega_1 \pm \omega_2$. In this manner, the waves with frequencies ω_1 and ω_2 excite waves with the sum and difference frequencies $\omega_\pm$. If the phase detuning

$$\Delta\mathbf{k} = \mathbf{k}_\pm - \mathbf{k}_1 \mp \mathbf{k}_2, \quad (3.118)$$

with $k_\pm = n(\omega_\pm)\omega_\pm/c$, is small, we can employ the general solution (3.18) to determine the slowly varying amplitudes $\mathbf{E}(\omega_\pm, z)$ of the excited waves. We orient the z axis along $\mathbf{k}_1 \mp \mathbf{k}_2$, while the interface between the medium and the vacuum lies perpendicular to the z axis. Next we examine the polarization wave propagating along the z axis into the medium. The vector $\mathbf{k}_\pm$ is also assumed to be directed along the z axis. In the field strength $\mathbf{E}_\pm(z, t)$ in the polarization wave we isolate a wave factor [cf. (3.17)] in the following manner:

$$\mathbf{E}_\pm(z, t) = \mathbf{E}(\omega_\pm, z)\exp(i\omega_\pm t - ik_\pm z), \quad (3.119)$$

with a slowly varying amplitude [cf. (3.18)]

$$E_i(\omega_\pm, z) = \sum_{jk} \chi_{ijk}^{(2)} E_j(\omega_1) E_k(\omega_2) 2\pi N \frac{k_\pm}{\Delta k_\pm}$$
$$\times [1 - \exp(i\,\Delta k_\pm z)]. \quad (3.120)$$

The excited waves will have a higher intensity when phase matching takes place, that is,

$$k_\pm = k_{1z} \pm k_{2z}. \quad (3.121)$$

For this condition to be satisfied, the vectors $\mathbf{k}_1$ and $\mathbf{k}_2$ of the incident waves must be directed at a certain angle θ to each other, that is, we must take into account the dispersion of the medium specified by (3.116).

Let us investigate the condition (3.121) using the simple example of the two frequencies being equal, $\omega_1 = \omega_2 = \omega$ and hence $k_1 = k_2 = k$, while the directions of $\mathbf{k}_1$ and $\mathbf{k}_2$ are different; in this respect the problem differs from the case of harmonic excitation (Section 3.2). Consider the excitation of the wave with the sum frequency 2ω (Figure 3.26). Since $k_1\cos(\theta/2) =$

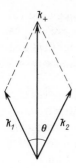

Figure 3.26. The phase-matching condition for the coupling of two waves with the same frequency.

$k_{\pm}/2$, condition (3.21) for this case takes the form

$$n_{2\omega} = n_{\omega}\cos(\theta/2).\qquad(3.122)$$

If we assume, as is common practice, that the refractive indices differ little from unity, then we find that $\theta \ll 1$, and, expanding $\cos(\theta/2)$ in a Taylor series ($\cos\alpha \approx 1 - \alpha^2/2$), we see that (3.122) yields

$$\theta = \left[8(n_{\omega} - n_{2\omega})\right]^{1/2}.\qquad(3.123)$$

If the angle between the wave vectors $\mathbf{k}_1$ and $\mathbf{k}_2$ of the coupled waves deviates from the value given by (3.123), the intensity of the wave with frequency 2ω will diminish as $\Delta k_{\pm}$, the phase detuning defined in (3.118), increases. Note that in this case the phase-matching condition is realized for anomalous dispersion, when $n_{\omega} > n_{2\omega}$.

In the general case of different frequencies $\omega_1 \neq \omega_2$, certain general conclusions can be drawn about the phase-matching condition (3.121). Assuming that the refractive indices differ little from unity, we see that phase matching requires that $\mathbf{k}_1$ and $\mathbf{k}_2$ be close in direction. In a medium without dispersion we conclude that the two vectors are collinear.

We see, therefore, that at fixed parameters of the nonlinear medium we can, by selecting the angle θ between the directions of the incident wave vectors, realize the matching condition and achieve excitation of a wave with a frequency that differs from the frequencies of the incident waves. Processes of this type, as already noted in the Introduction to this book, are called coherent (or parametric) (Akhmanov and Khokhlov, 1966), since the intensity of the excited wave depends essentially on the phase detuning, which in this case is determined by a parameter, the angle θ between the directions of propagation of the incident waves; the terminology appeared by analogy with the phenomenon of excitation of parametric oscillations in electronics (Louisell, 1960). Obviously, when phase matching is achieved for three interacting waves, energy transfer from the incident wave to the excited wave becomes possible. The operation of parametric amplifiers and generators of coherent radiation is based on this possibility (Akhmanov and Khokhlov, 1966; Dmitriev and Tarasov, 1982).

The case of an arbitrary number of coupled electromagnetic waves can be considered in a similar way. The number of frequencies at which waves can be excited increases sharply with the number of external electromagnetic waves incident on the medium. The majority of interaction processes between coupled waves are parametric, since phase matching can be achieved only by selecting the parameters that characterize the incident waves.

Up till now we have studied the coupling between waves determined by quadratic polarization. Processes described by cubic and higher-order polarizations can be considered in a similar manner. Since the topic of our discussion is an atomic gas, let us turn to processes determined by cubic polarization. The number of different interactions in this case is very extensive.

3.4.3. Four-Wave Coupling

A natural generalization for the process considered in Section 3.4.2 is the case where three different electromagnetic waves with frequencies ω_1, ω_2, and ω_3 are incident on a nonlinear medium and excite a fourth wave in it. As a result of the interaction of these waves, there appears a nonlinear polarization wave with a frequency $\nu = \pm\omega_1 \pm \omega_2 \pm \omega_3$.

A typical Feynman diagram for the cubic susceptibility $\chi^{(3)}(\nu; \omega_1, \omega_2, \omega_3)$ in the four-wave interaction is depicted in Figure 3.27. In this process energy conservation holds: $\nu = \omega_1 + \omega_2 + \omega_3$. The coherence of the above process stems from the fact that for a high probability of this process the law of conservation of momentum in the interaction of the four photons with each other must be satisfied either exactly or approximately: $\mathbf{k}_\nu = \mathbf{k}_1 + \mathbf{k}_2 + \mathbf{k}_3$.

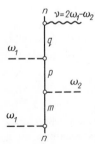

Figure 3.27. A Feynman diagram for the cubic susceptibility describing coherent anti-Stokes Raman scattering as a particular case of four-wave coupling.

There are many processes whose Feynman diagrams differ from the one depicted in Figure 3.27 either by the order in which the photons are absorbed and emitted or by the interchange of absorption and of emission. Below we will provide a detailed discussion of the main types of such processes and will give the three most common examples of four-wave coupling.

Of course, far from resonances the Feynman diagram in Figure 3.27 does not in any way stand out against the other numerous diagrams for the cubic susceptibility. Besides since the field strengths in the incident waves are low compared to the atomic fields, the value of the corresponding nonlinear polarization is negligible. For this reason, in which follows we will speak only of processes that stand out due to resonances in the intermediate states. Since we remain within the scope of the unperturbed-field approximation, we will not consider situations in which the fields are so high that in the vicinity of resonances they affect the resonance behavior of the nonlinear susceptibility. Thus, broadening of the emerging resonances is due to standard mechanisms, such as spontaneous radiation broadening, collision (impact) broadening, collision quasistatic broadening, Doppler broadening, and the like.

In reality two or even one instead of three external electromagnetic waves are usually incident on the nonlinear medium. For example, if the field with ω_1 is high, then in the above scheme $\omega_3 = \omega_1$. When there is only one incident wave, the role of the fields with ω_2 may be played by the stimulated-Raman-scattering wave generated in the nonlinear medium.

3.4.4. Coherent Anti-Stokes Raman Scattering

Let us take the particular case of the Feynman diagram in Figure 2.12, where $\omega_3 = \omega_1$ and a photon with frequency ω_2 is emitted in a stimulated manner, so that in the medium a nonlinear polarization wave with a frequency of $\nu = 2\omega_1 - \omega_2$ (Figure 3.27) is excited. Such a process is known as *coherent anti-Stokes Raman scattering* (see Akhmanov and Koroteev, 1981, and Section 3.3.5). Indeed, since n is the ground state, $\omega_{pn} > 0$. In conditions of a two-photon resonance transition from n to p we have $\omega_{pn} \approx \omega_1 - \omega_2$. Thus, for the frequency ν of the excited polarization wave we have $\nu \approx \omega_1 + \omega_{pn} > \omega_1 > \omega_2$. So the anti-Stokes frequency $\nu = 2\omega_1 - \omega_2$ proves to be higher than the frequency ω_1 of the incident radiation (and higher than ω_2). Hence, the given process may be used to excite coherent ultraviolet radiation by using coherent radiation in the visible range.

As noted earlier, the field with frequency ω_2 is not necessarily an external field. It may be a component in stimulated Raman scattering of a

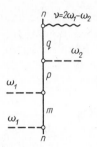

Figure 3.28. The Feynman diagram for the cubic susceptibility in coherent anti-Stokes Raman scattering under two-photon absorption of the wave with frequency ω_1.

field of frequency ω_1 incident on the nonlinear medium. This case has been discussed in Section 3.3.5. Note that the relationship $\omega_{pn} \approx \omega_1 - \omega_2$ becomes exact, and state p becomes populated.

The coherence of this process manifests itself in the fact that the probability is determined by the size of the phase detuning $\Delta \mathbf{k} = \mathbf{k}_\nu - (2\mathbf{k}_1 - \mathbf{k}_2)$, just as in the case of third-harmonic excitation.

Coherent anti-Stokes Raman scattering is possible also under two-photon absorption of the radiation with frequency ω_1. The Feynman diagram for the corresponding cubic susceptibility is shown in Figure 3.28. The condition of two-photon resonance has the obvious form $2\omega_1 \approx \omega_{pn}$.

As an example of such a scheme one can take the results obtained by Hartig, 1978 (Figure 3.29). Intense radiation of a dye laser acted on a nonlinear medium of sodium vapor. The frequency of this field, ω_1 [corresponding to a wavelength $\lambda(\omega_1) = 578$ nm], was tuned to the two-photon resonance between the ground state $n = 3S_{1/2}$ of a sodium atom and the excited state $p = 4D_{5/2}$, so that $\omega_{pn} \approx 2\omega_1$. A field of frequency $\omega_2 \approx 2\omega_1 - \omega_{qn}$ was not external, but was excited in the medium (in stimulated hyper-Raman scattering) as a polarization wave, with $q = 4P_{3/2}$. Since $2\omega_1 \approx \omega_{pn}$, we see that $\omega_2 \approx \omega_{pq}$ (Figure 3.29). This field constituted infrared radiation with a wavelength $\lambda(\omega_2) = 2.34$ μm. The resulting polarization wave with a frequency $\nu = 2\omega_1 - \omega_2 \approx \omega_{qn}$ (which corre-

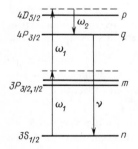

Figure 3.29. Emission-absorption diagram for coherent anti-Stokes Raman scattering in sodium vapor (Hartig, 1978).

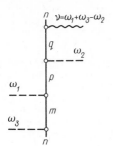

Figure 3.30. The Feynman diagram for the cubic susceptibility describing coherent anti-Stokes Raman scattering under two-photon absorption, $\omega_1 + \omega_3 \approx \omega_{pn}$ (Lumpkin, 1978).

sponded to a wavelength of 330 nm) lay in the ultraviolet region of the spectrum.

The findings of Lumpkin, 1978, can be cited as an example in which two external fields were used. The researcher used the radiation of a ruby laser at $\omega_1 = 14{,}277$ cm^{-1} and the first Stokes component at $\omega_3 = 13{,}044$ cm^{-1}. The two waves were incident on a medium of potassium vapor. The excitation of atomic states and the absorption and emission of photons are shown in Figure 3.30 by employing the Feynman diagram for the corresponding cubic susceptibility. The sum frequency $\omega_1 + \omega_3$ of the two-photon excitation was in resonance with the frequency of the atomic transition $n = 4S_{1/2} \rightarrow p = 6S_{1/2}$. A field of frequency $\omega_2 = \omega_1 + \omega_3 - \omega_{qn}$, with $q = 5P_{3/2}$ [the wavelength $\lambda(\omega_2) = 3.6$ μm], was the polarization wave of stimulated hyper-Raman scattering in the nonlinear medium. As a result a polarization wave in the violet spectral range with a frequency $\nu = \omega_1 + \omega_3 - \omega_2$ [the wavelength $\lambda(\nu) \approx 400$ nm] was excited. The possible mechanisms of population of states $m = 4P_{1/2}$ and $m = 4P_{3/2}$ of a potassium atom when they are resonantly excited by a field of frequency ω_3 (Figure 3.30) have been discussed by Anikin, Kryuchkov, and Ogluzdin, 1975.

3.4.5. Coherent Two-Photon Radiation

Here is another example of the excitation of a nonlinear polarization wave with frequency ν. The corresponding Feynman diagram for the nonlinear susceptibility $\chi^{(3)}(\nu; \omega_1, -\omega_2, -\omega_3)$ is shown in Figure 3.31. Energy conservation yields the following expression: $\nu = \omega_1 - \omega_2 - \omega_3$. The value of $\chi^{(3)}$ is appreciable only when the two-photon resonance condition is met: $\omega_{pn} \approx \omega_1 - \omega_3$. The frequency ν of the polarization wave then takes the form $\nu \approx \omega_{pn} - \omega_2$. One photon of frequency ω_1 is absorbed and three photons of frequencies ω_2, ω_3, and ν are emitted in this process.

The wave with frequency ω_3 may not be an external wave but a wave created in the process of stimulated Raman scattering when a strong wave of frequency ω_1 is incident on the nonlinear medium. Then to ensure that

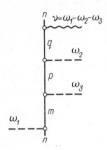

Figure 3.31. The Feynman diagram for the cubic susceptibility describing two-photon emission under the conditions of the experiment conducted by Bethune, Lankard, and Sorokin, 1979.

$\omega_3 \approx \omega_1 - \omega_{pn}$ it is not necessary to select the frequency ω_1. Since ω_{pn} is positive, $\omega_3 < \omega_1$, that is, ω_3 is a Stokes frequency. The frequency of the nonlinear polarization wave in this case, $\nu \approx \omega_{pn} - \omega_2$, is independent of the frequency ω_1 of an incident wave. Note that here not three but two waves are incident on the nonlinear medium, waves with frequencies ω_1 and ω_2. To amplify the excitation of the stimulated-Raman scattering wave with frequency ω_3, the frequency ω_1 is usually chosen so that it is in resonance with that of one of the atomic transitions, ω_{mn}, that is, $\omega_1 \approx \omega_{mn}$.

To illustrate this scheme, we refer to the work of Bethune, Lankard, and Sorokin, 1979. The vapor of atomic potassium was taken as the nonlinear medium in the experiment, with the ground state $n = 4S_{1/2}$ and the excited state $m = 5P_{3/2}$ (Figure 3.32). The frequency $\omega_1 \approx \omega_{mn}$ was selected so as to ensure the excitation of an intense resonance wave of stimulated Raman scattering with a frequency $\omega_3 = \omega_1 - \omega_{pn}$, where $p = 5S_{1/2}$. Finally, $\nu = \omega_{pn} - \omega_2 < \omega_{pn} < \omega_1$, where ω_2 is the frequency of the second incident wave. This scheme clearly shows that ν is a Stokes frequency, so that the excitation of a nonlinear polarization wave takes place in the infrared frequency range, in contrast to coherent anti-Stokes Raman scattering considered in Section 3.4.4. As Figure 3.32 shows, in the case at hand the difference combination Stokes frequency is excited during resonant stimulated Raman scattering, and the interaction of the waves produces three photons out of one, that is, two photons are emitted. The given process, therefore, can be named *coherent two-photon emission*.

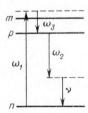

Figure 3.32. Emission-absorption diagram for the process represented by the Feynman diagram shown in Figure 3.31.

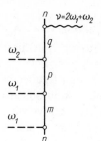

Figure 3.33. The Feynman diagram for the cubic susceptibility corresponding to the excitation of the wave with the sum frequency $2\omega_1 + \omega_2$ in sodium vapor (Bloom, Yardley, Young, and Harris, 1974).

3.4.6. Excitation of the Wave with the Sum Frequency

The Feynman diagram for the nonlinear susceptibility $\chi^{(3)}(\nu; \omega_1, \omega_1, \omega_2)$ depicted in Figure 3.33 illustrates the excitation of the wave with the sum frequency $\nu = 2\omega_1 + \omega_2$ as a result of four-wave coupling. Two waves with frequencies ω_1 and ω_2 are incident on the nonlinear medium. When $\omega_1 = \omega_2$, the process reduces to third-harmonic excitation. As in third-harmonic excitation, the excitation of the wave with the sum frequency is coherent, that is, its probability is determined by the phase detuning $\Delta k = k_\nu - 2k_1 - k_2$.

The most interesting possibility here is that of a two-photon resonance, $2\omega_1 \approx \omega_{pn}$. The state p becomes populated, that is, two-photon absorption of the incident wave of frequency ω_1 takes place (for details see Section 3.7).

An experiment in excitation of the sum frequency $\nu = 2\omega_1 + \omega_2$ in sodium vapor has been conducted by Bloom, Yardley, Young, and Harris, 1974. Note that the experiment had nothing to do with the coupling of the wave ω_2 with the second-harmonic wave $2\omega_1$ that appears in the propagation of the wave with frequency ω_1 through the medium, since, as noted earlier, second-harmonic excitation is forbidden by parity conservation. The experiment, whose level diagram is shown in Figure 3.34, realized the resonance between the frequency $2\omega_1$ and the frequency of the transition

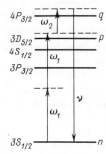

Figure 3.34. Emission-absorption diagram for the excitation of the wave with the sum frequency $2\omega_1 + \omega_2$ in sodium vapor, corresponding to the Feynman diagram shown in Figure 3.33.

from the ground state $n = 3S_{1/2}$ to the state $p = 3D_{5/2}$. To ensure an additional resonant increase in the nonlinear susceptibility $\chi^{(3)}$, the frequency ω_2 was chosen close to the frequency of the atomic transition between states $q = 4P_{3/2}$ and $3D_{5/2}$.

Since three photons are absorbed in the excitation of the sum frequency (Figure 3.34), state q proves to be highly excited if resonance conditions are met, so that ionization from this state is a channel strongly competitive with the process in which the atom returns to the initial ground state n and a high-energy photon of frequency $\nu = 2\omega_1 + \omega_2$ is emitted. One method of suppressing ionization has been suggested by Crance and Armstrong, 1982. An autoionization state was suggested for the role of q, under the conditions for the realization of this scheme and the resonance conditions $2\omega_1 \approx \omega_{pn}$ and $\omega_2 \approx \omega_{qp}$ (for a similar approach in discussing the competitiveness of ionization in relation to third-harmonic excitation see Section 3.2.6). The field with frequency ω_2 was found to have a marked influence on the probability of autoionization decay of this state. This effect is determined not only by the intensity of the radiation with frequency ω_2 but also by the difference between ω_2 and the atomic transition frequency ω_{qp}. At certain intensities and resonance misfits, the autoionization probability, as calculations conducted by Crance and Armstrong, 1982, have shown, may be strongly suppressed. This, in turn, increases the probability of excitation of the sum frequency $\nu = 2\omega_1 + \omega_2$.

The possibility of using autoionization-like quasi energy states, which appear in the continuous spectrum when an external field is applied, to increase the effectiveness of the process of frequency mixing has been studied both theoretically and experimentally by Dimov, Metchkov, Pavlov, and Stamenov, 1982 (for more details see Section 3.2.6 as well as Dimov, Pavlov, Geller, and Popov, 1983; Geller and Popov, 1981; and Pavlov, Dimov, Metchkov, Mileva, and Stamenov, 1982, where third-harmonic excitation is examined).

Finally, we would like to note that multiphoton frequency mixing also offers certain perspectives. As an example we cite the work by Drabovich, Ignatovichyus, Kupris, Matsulyavichyus, Pershin, Synyavskiĭ, Smil'gyavichyus, and Surovegin, 1982, who observed six- and eight-photon resonance mixing of frequencies in sodium vapor. Ways of ensuring that phase matching conditions are met in such processes have been discussed by Le Boitex, Raj, Gao, Bloch, and Ducloy, 1984.

3.4.7. The Manley–Rowe Law and the Phase-Matching Conditions in Four-Wave Coupling

As has been seen, in the majority of realistic four-wave resonance coupling schemes, one of the waves is the first component in the stimulated-Raman-scattering spectrum (e.g. see Figure 3.27). Such coupling, therefore, can be called neither coherent nor incoherent. Indeed, while phase matching (Section 3.3) is unnecessary for exciting the first stimulated-Raman-scattering component with frequency ω_2 (Figure 3.27), it is certainly necessary for exciting a polarization wave with frequency $\nu = 2\omega_1 - \omega_2$. Here ω_1 is the frequency of the incident wave. In ideal conditions, phase matching $\Delta\mathbf{k} = 0$ is achieved when the wave vectors of waves 1, 2, 3, and 4 are collinear; for example, for the process described by the Feynman diagram in Figure 3.33 the phase-matching condition is

$$[2\omega_1 + \omega_2]\,\Delta n(2\omega_1 + \omega_2) = 2\omega_1\,\Delta n(\omega_1) + \omega_2\,\Delta n(\omega_2), \quad (3.124)$$

where

$$\Delta n(\omega_i) = 2\pi N \chi^{(1)}(\omega_i; -\omega_i) \quad (3.125)$$

is the linear variation of the refractive index at ω_i in the gaseous medium, and N is the number of atoms per unit volume of gas. If in (3.124) one of the frequencies is low (say, if the frequency ω_2 lies in the far infrared), then for all practical purposes it can be ignored.

As we have seen, a specific feature of four-photon resonance coupling is that the intermediate state p becomes populated in the excitation of a stimulated-Raman-scattering wave. If we turn, say, to the excitation scheme shown in Figure 3.31, we can say that the absorption of one photon of frequency ω_1 leads to the emission of a photon of frequency ω_3, that is, the total number of photons with frequencies $\omega_1\omega_3$ remains unchanged:

$$N(\omega_1) + N(\omega_3) = \text{const.} \quad (3.126)$$

This fact can be expressed in terms of radiation intensities at ω_1 and ω_3:

$$\frac{I(\omega_1)}{\omega_1} + \frac{I(\omega_3)}{\omega_3} = \text{const} = \frac{I_0(\omega_1)}{\omega_1}, \quad (3.127)$$

where $I_0(\omega_1)$ is the intensity of the wave with frequency ω_1 in the left half space $z < 0$, at the entrance into the nonlinear medium. Formulas of the type (3.127) are known as the *Manley–Rowe law* (Manley and Rowe, 1959).

A similar Manley–Rowe law for the polarization wave with frequency ν within the scope of the same scheme (Figure 3.31) has the form

$$\frac{I(\omega_2)}{\omega_2} - \frac{I(\nu)}{\nu} = \text{const} = \frac{I_0(\omega_2)}{\omega_2}, \quad (3.128)$$

where the minus reflects the fact that the two photons of frequencies ω_2 and ν are emitted in a single act of four-photon coupling (Figure 3.31), that is, the variation in the number of photons of frequency ν in the medium is equal to the variation in the number of photons of frequency ω_2.

In a similar manner the Manley–Rowe law can be written for other cases of four-photon coupling discussed above, where the intermediate state p is populated thanks to stimulated Raman scattering, which leads to the separation in time of the absorption of a photon of frequency ω_1 and the emission of a photon of frequency ω_3 from the emission of photons of frequencies ω_2 and ν.

Essentially, the Manley–Rowe law reflects the conservation of the number of photons taking part in the coupling process. The practical value of this law is obvious, since it enables us to determine the intensity of the excited radiation if we know the frequencies of the interacting waves and the intensity of the exciting radiation.

3.4.8. Tensor Properties of the Cubic Susceptibility for Four-Wave Coupling in Alkali-Atom Vapors

The general expression for the tensor $\chi^{(3)}_{ijkl}$ averaged over the directions of the angular momentum of an atom freely oriented in space (that is, over the magnetic quantum numbers) is represented in the form of three products of Kronecker deltas, $\delta_{ij}\delta_{kl}, \delta_{ik}\delta_{jl}, \delta_{il}\delta_{jk}$ (for more details see Section 3.6.1), since there are no selected directions in space. Instead of these three products it is convenient to introduce three other linear combinations of the above products [e.g. see Rapoport, Zon, and Manakov, 1978, Eq. (4.10)]. The new combination are called, respectively, the *scalar*, *symmetric*, and *antisymmetric parts* of nonlinear susceptibility $\chi^{(3)}_{ijkl}$ and are given below.

If we turn our attention to the Feynman diagrams for the cubic susceptibilities considered in this section, we will see that for alkali atoms the initial state n in these diagrams is always the S state, the ground state of the valence electron in the atom. But then, according to dipole selection rules, the intermediate state p, excited as a result of absorption of two photons, is either again an S state (but with a different principal quantum number) or a D state.

As can be shown (see Il'inskiĭ and Taranukhin, 1976), when the intermediate state p has an orbital angular momentum $l = 0$ (i.e. is an S state), only the scalar part $A\delta_{ij}\delta_{kl}$ of the cubic susceptibility tensor $\chi^{(3)}_{ijkl}$ is nonzero. The selection rules for this part are the same as the selection rules for the matrix elements of a scalar quantity. But if the intermediate state p has an orbital angular momentum $l = 2$ (i.e. is a D state), there is also a symmetric part in the tensor $\chi_{ijkl}^{(3)}$:

$$\chi_{ijkl}^{(3)} = B\left[\delta_{ik}\,\delta_{jl} + \delta_{il}\,\delta_{jk} - \tfrac{2}{3}\delta_{ij}\,\delta_{kl}\right], \tag{3.129}$$

with selection rules that coincide with those for the matrix elements of electric quadrupole radiation (Berestetskiĭ, Lifshitz, and Pitaevskiĭ, 1980, §60). The antisymmetric part $C[\delta_{ik}\,\delta_{jl} - \delta_{il}\,\delta_{kj}]$, which formally contributes to $\chi_{ijkl}^{(3)}$, disappears completely. The selection rules for this part coincide with those for the matrix elements of magnetic dipole radiation (Berestetskiĭ, Lifshitz, and Pitaevskiĭ, 1980, §60).

As for the competitiveness of the various components of the doublets of excited states of alkali atoms, according to Bethe and Salpeter, 1957, the largest matrix elements correspond to the transition $n = S_{1/2} \to m = P_{3/2} \to p = D_{5/2}$ and the one into state $p = S_{1/2}$. The total angular momentum changes in the same direction as the orbital angular momentum.

3.4.9. Conclusion

In this section we have discussed the most general case when several waves with different frequencies and different directions of propagation are incident on the nonlinear medium. The main and highly remarkable conclusion that follows from a study of this case is that at a certain ratio between the frequencies and wave vectors of the incident waves, new waves may be effectively excited in the medium. The excited waves appear to be coupled with the exciting waves, and this makes possible an effective energy transfer between these waves. At the base of the process of generation of coupled waves lies the nonlinear polarization of the medium, which is caused by the overall effect that all the incident waves have on the medium.

This phenomenon of wave coupling considerably broadens the spectrum of frequencies at which it is possible to create intense coherent radiation by means of nonlinear-optics processes, using the radiation of high-power lasers as the exciting radiation. In contrast to the above-discussed processes of excitation of optical harmonics and of excitation of components of stimulated Raman scattering, which make it possible to arrive at a discrete set of frequencies in the new waves, wave coupling in a nonlinear medium enable one to create waves with smoothly varying frequencies.

The great practical importance of wave coupling is well known. We note only the two areas most broadly developed: parametric generators of coherent radiation (Dmitriev and Tarasov, 1982) and coherent anti-Stokes Raman-scattering spectroscopy (Akhmanov and Koroteev, 1981).

This section closes the discussion of nonlinear transformation of the frequency of light, since the three processes discussed—higher-harmonic excitation, stimulated Raman scattering, and coupled waves—practically exhaust all phenomena in this field. In our further exposition we turn to questions pertaining to changes in the direction of propagation of light waves in the nonlinear interaction with a medium, changes in light polarization, and nonlinear absorption of light.

3.5. NONLINEAR REFRACTION

3.5.1. Introduction

The picture of reflection and refraction of light of low intensity at the interface between a medium and vacuum or at the interface between two media with different refractive indices is well known (see Sections 1.3.5 and 1.5). But when the incident light wave is intense and spatially inhomogeneous, quite different phenomena emerge. One is the nonlinear refraction of light, that is, the change in the direction of its propagation. Since the nonlinearity of the medium is induced (that is, emerges as a result of the interaction of the strong light field and the medium), the medium becomes inhomogeneous in a wave with an amplitude inhomogeneously distributed over the wavefront. This means that at different points in space the averaged optical characteristics of the medium (e.g., the refractive index) differ. From linear optics it is well known that the light does not propagate along a straight line in a medium with a variable refractive index.

Note that the model considered here, an incident wave with an inhomogeneous distribution of the amplitude over the wavefront, is quite realistic. Indeed, the laser beam is always restricted spatially, while the amplitude varies over the wave from the maximal value on the axis to zero on the periphery.

A particular case of nonlinear refraction is the self-focusing of laser radiation, a phenomenon extensively studied both theoretically and experimentally (Akhmanov, Khokhlov, and Sukhorukov, 1972; Akhmanov, Sukhorukov, and Khokhlov, 1968; Lugovoĭ and Prokhorov, 1974; Marburger, 1975; Shen, 1975a, b). Self-focusing, which occurs along with other nonlinear phenomena, drastically changes the properties of the incident radiation; and hence it greatly influences the development of other nonlinear effects and makes it difficult to analyze the experimental data.

In this section we will consider the basics of nonlinear refraction within the same general framework in which we have been studying all processes of nonlinear interaction of strong light with atomic gases. At first glance it may seem that since the very essence of nonlinear refraction consists in the variation of the spatial distribution of radiation in a wave propagating in a medium, the unperturbed-field approximation cannot be employed to describe this phenomenon. But in reality this is not so. The unperturbed-field approximation can be used if we restrict our discussion to small angles of deviation from the direction of propagation of the incident wave. This greatly simplifies, as in other cases, the mathematical description of nonlinear refraction. As usual, we will also take up the case of steady-state interaction of the radiation and the medium. And at the end of this section we will briefly discuss the most important phenomena that fall outside the framework of the unperturbed-field approximation.

3.5.2. The Phenomenon of Nonlinear Refraction

Let us consider the physical meaning of nonlinear refraction, using a simple example: a plane wave whose field strength depends on the transverse coordinate y propagates in a vacuum ($z < 0$) along the z axis and falls onto a gaseous medium ($z \geq 0$) normally to the planar interface at $z = 0$. Of course, in a real laser beam there is axial symmetry, and the intensity of the radiation falls from the axis to the periphery by the Gaussian law. In addition, the wavefront is not planar and the wave diverges. However, the main laws can be formulated using the simple model of Section 1.5. The role of the real geometry of laser beams will be discussed at the end of this section.

In the polarization $\mathbf{P}(y, z, t)$ that appears in the medium under the action of the incident wave there are terms that are linear and nonlinear in the electric field strength $\mathbf{E}(y, z, t)$. The proportionality factor between $\mathbf{P}$ and $\mathbf{E}$ in the linear term is $4\pi N\chi^{(1)}$, which is independent of y (Section 1.5.4). Correspondingly, as established in Section 1.5, there is no linear refraction of the light beam. The dependence on y manifests itself only in the nonlinear terms in the expansion of the polarization vector in powers of the field strength.

The following discussion is carried out within the framework of the unperturbed-field approximation for the incident wave (Section 3.1), as we have done in previous cases. Thus, as in Section 1.5, we will, strictly speaking, consider only small deviations from the direction of propagation of the incident wave. In contrast to the linear case, we will also restrict our discussion to small distances z from the interface. The concrete limitations will be discussed below. We will also discuss what happens when the unperturbed-field approximation does not work.

Let us see how nonlinear refraction emerges by using the simplest dependence (1.85) of the field strength $\mathbf{E}$ on y, which satisfies the Maxwell equations in a vacuum and has been discussed in detail in Section 1.5.4:

$$\mathbf{E}(y, z, t) = \mathbf{E}_\omega\left(1 - \frac{y}{a}\right)\exp(i\omega t - ikz). \tag{3.130}$$

This constitutes a linearly polarized wave propagating along the z axis and having a frequency ω, a wave number $k = \omega/c$, and a vector $\mathbf{E}_\omega$ directed along the x axis. The quantity a characterizes the transverse size of the inhomogeneity of the wave. This inhomogeneity is considered small, so that the deviations y from the beam's axis ($y = 0$) are small compared to a. In addition, a is assumed to be so large compared to the wavelength of the radiation that $ka \gg 1$. We will use this inequality below.

Now let us turn to the nonlinear part in the susceptibility, assuming for the sake of simplicity that there is no linear part (e.g. $\chi^{(1)}$ is very small). Since we assume that the medium is an atomic gas, the first nonlinear term in the susceptibility is the cubic. We will restrict our discussion to this term and ignore all higher-order terms in the expansion of the susceptibility. Among the various Feynman diagrams contributing to the cubic susceptibility we must retain only those corresponding to the polarization wave at $\nu = \omega$, that is, at the frequency of the incident wave. One diagram of this type is depicted in Figure 2.9.

Let us denote the corresponding nonlinear atomic susceptibility by $\chi^{(3)} = \chi^{(3)}(\omega; \omega, -\omega, \omega)$. In the unperturbed-field approximation, the nonlinear polarization has the form [according to (3.2)]

$$\mathbf{P}^{(3)} = \chi^{(3)}\mathbf{E}_\omega^3\left(1 - \frac{y}{a}\right)^3 \exp(i\omega t - ikz). \tag{3.131}$$

Both vectors, $\mathbf{P}^{(3)}$ and $\mathbf{E}_\omega$, are directed along the x axis. The Maxwell equation for the field strength in the weak polarization wave $\delta\mathbf{E}(y, z, t)$ at frequency ω in the unperturbed-field approximation is obtained by a method similar to that used in deriving Eq. (1.86). The result is

$$\frac{\partial^2 \delta\mathbf{E}(y, z, t)}{\partial y^2} + \frac{\partial^2 \delta\mathbf{E}(y, z, t)}{\partial z^2} + k^2 \delta\mathbf{E}(y, z, t)$$

$$= -4\pi Nk^2\chi^{(3)}\mathbf{E}_\omega^3\left(1 - \frac{y}{a}\right)^3 \exp(i\omega t - ikz). \tag{3.132}$$

Isolating the main part in the wave dependence along the z axis, that is, writing

$$\delta\mathbf{E}(y, z, t) = \delta\mathbf{E}_\omega(y, z)\exp(i\omega t - ikz), \tag{3.133}$$

and substituting this into Eq. (3.132), we arrive at the following equation for the amplitude of the polarization wave:

$$\left(\frac{\partial^2}{\partial y^2} + \frac{\partial^2}{\partial z^2} - 2ik\frac{\partial}{\partial z}\right)\delta\mathbf{E}_\omega(y, z) = -4\pi Nk^2\chi^{(3)}\mathbf{E}_\omega^3\left(1 - \frac{y}{a}\right)^3. \tag{3.134}$$

Since $\delta E_\omega = 0$ at $z = 0$, we look for the solution to Eq. (3.134) in the form [see also (1.87)]

$$\delta E_\omega(y, z) = \left[A(y)z + Bz^2\right]\left(1 - \frac{y}{a}\right)E_\omega. \qquad (3.135)$$

Substituting (3.135) into Eq. (3.134) and neglecting the derivative $\partial^2/\partial z^2$ (below we will find the conditions when this is justified) yields the sought solution:

$$A(y) = -2\pi i k N \chi^{(3)}\left(1 - \frac{y}{a}\right)^2 E_\omega^2, \qquad B = -3\pi N \chi^{(3)}\left(\frac{E_\omega}{a}\right)^2. \qquad (3.136)$$

The term with $\partial^2/\partial z^2$ in (3.134) can be neglected if $kz \gg 1$ and $ka \gg 1$, conditions we assume to be met. The approximation in which the term with $\partial^2/\partial z^2$ is neglected in the Maxwell equations is known as the *truncated-Maxwell-equations approximation* (Butylkin, Kaplan, Khronopulo, and Yakubovich, 1984).

Substituting (3.136) into (3.135) and the result into (3.133), we arrive at the following expression for the field strength in the nonlinear polarization wave [cf. the solution (1.87) for the weak linear polarization wave]:

$$\delta E(y, z, t) = -\pi N \chi^{(3)}\left(1 - \frac{y}{a}\right)E_\omega^3 \exp(i\omega t - ikz)$$

$$\times\left[2ikz\left(1 - \frac{y}{a}\right)^2 + 3\left(\frac{z}{a}\right)^2\right]. \qquad (3.137)$$

Adding the field strength (3.137) in the polarization wave to the field strength (3.130) in the incident wave, we obtain the total field strength in the wave in the medium [cf. (1.89)]:

$$E' = E + \delta E = E_\omega\left(1 - \frac{y}{a}\right)\exp(i\omega t - ikz)$$

$$\times\left[1 - 2\pi i k z N \chi^{(3)}E_\omega^2\left(1 - \frac{y}{a}\right)^2 - 3\left(\frac{z}{a}\right)^2\pi N \chi^{(3)}E_\omega^2\right]. \qquad (3.138)$$

Note that the incident wave with the field strength (3.130) is assumed to be weakly inhomogeneous, that is, a is much greater than the characteristic values of y. In (3.138) we separate the terms in the square brackets that depend on y from those that do not: $(1 - y/a)^2 \approx 1 - 2y/a$ [i.e., we neglect the small quantity $(y/a)^2 \gg 1$]. Combining this with (3.130), we can rewrite the solution (3.138) approximately in the following manner:

$$\mathbf{E}'(y, z, t) = \mathbf{E}(y, z, t)\left\{1 - 2\pi i k z N \chi^{(3)} E_\omega^2\right.$$
$$\left. - 3\left(\frac{z}{a}\right)^2 \pi N \chi^{(3)} E_\omega^2 + 4\pi i k z N \chi^{(3)}\left(\frac{y}{a}\right) E_\omega^2\right\}. \quad (3.139)$$

This solution satisfies Maxwell's equation (3.134) only approximately. Let us write the expression in the braces in exponential form [similar to the way this was done when from (1.47) and (1.88) we went over to (1.48) and (1.89), respectively]:

$$\mathbf{E}'(y, z, t) = \mathbf{E}_\omega \exp\left(-ik_z'z - ik_y y\right)$$
$$\times \left[1 - \frac{y}{a} - 3\left(\frac{y}{a}\right)^2 \pi N \chi^{(3)} E_\omega^2\right], \quad (3.140)$$

where

$$k_z' = \left[1 + 4\pi N \chi^{(3)} E_\omega^2\right]^{1/2} k \quad (3.141)$$

is the wave number along the z axis, which has changed because of nonlinear polarization [see also (3.51)], and

$$k_y = -4\pi k\left(\frac{z}{a}\right) N_\chi^{(3)} E_\omega^2 \quad (3.142)$$

is the wave number along the y axis, which has changed because of nonlinear polarization and the presence of transverse inhomogeneity in the wave in the form of the parameter a. The solution (3.140) also demonstrates that the profile of the amplitude of the field strength undergoes a small change in the medium.

Thus, the wave in the medium propagates along a direction that differs from that for the incident light wave, in which $k_y = 0$. Note also that in contrast to k_z', which does not vary with z, the wave number k_y grows with z, according to (3.142).

As in the linear case discussed in Section 1.5, here we assume that the refraction of the light wave is weak. From (3.141) it also follows that $\Delta k_z = k_z' - k_z \ll k$ yields

$$N\chi^{(3)}E_\omega^2 \ll 1. \quad (3.143)$$

This inequality can be called the *condition of weak nonlinearity* of the medium.

What are the conditions of applicability of the solution (3.140) for the field strength $\mathbf{E}'$ in the light wave in the medium? While the transitions from (1.47) to (1.48) and from (1.88) to (1.89) could be considered exact in view of the linearity of the Maxwell equations, a similar transition from (3.139) to (3.140) is, of course, only approximate because of the nonlinearity of the polarization in the field strength. To determine the degree to which such a transition is considered approximate, we substitute the solution (3.140) to the nonlinear Maxwell equation for the field strength $\mathbf{E}'$ in the medium, written without regard for the unperturbed-field approximation:

$$\frac{\partial^2 \mathbf{E}'}{\partial y^2} + \frac{\partial^2 \mathbf{E}'}{\partial z^2} + k^2 \mathbf{E}' = -4\pi N k^2 \chi^{(3)} |\mathbf{E}'|^2 \mathbf{E}'. \qquad (3.144)$$

Substituting (3.140) into (3.144) yields an identity if

$$k_y y \ll 1 \qquad (3.145)$$

(of course, the identity is satisfied only qualitatively). If the condition (3.145) is met, we can expand the exponential in (3.140) in powers of $k_y y$ and, retaining only the linear term, arrive at (3.139). In Section 3.5.5 we will discuss when this inequality is valid.

3.5.3. Propagation of a Light Ray in a Nonlinear Medium

In terms of the geometrical optics of light rays, the expression (3.140) represents a light ray refracted in plane in the yz plane. The angle of inclination of this ray to the z axis at the point with coordinates (y, z) is

$$\alpha \approx k_y / k_z = -4\pi \frac{z}{a} N \chi^{(3)} E_\omega^2. \qquad (3.146)$$

This formula is applicable when α is much less than unity, or the angle is small. Thus, k_y must be much less than k. Note that in (3.146) we ignored the small addition Δk_z to the wave number k_z because the condition (3.143) of weak nonlinearity of the medium is met.

We note that α does not remain constant but increases as z varies, with the result that the path of the ray is not a straight line. As (3.146) shows, the direction in which the ray is deflected depends on the sign of the nonlinear susceptibility of $\chi^{(3)}$.

Since in geometrical optics the inclination angle α is equal to the derivative dy/dz, Eq. (3.146) yields the following equation for the trajectory of the ray in differential form:

$$\frac{dy}{dz} = -4\pi\frac{z}{a}N\chi^{(3)}E_\omega^2.$$ (3.147)

Separating the variables and integrating with $y(0) = y_0 \ll a$, where y_0 is the initial coordinate of the ray along the y axis when the ray enters the medium, we arrive at the equation for the trajectory of the ray in explicit form:

$$y_0 - y = 2\pi\frac{z^2}{a}N\chi^{(3)}E_\omega^2.$$ (3.148)

From (3.148) it follows that the trajectory of the ray is a parabola. The condition $y_0 \ll a$ ensures the validity of the condition of the weak inhomogeneity of the light beam, $y \ll a$, which we imposed earlier, since we are interested in the region where $y \sim y_0$.

Up till now we have neglected the linear term in the polarization of the medium. If it is taken into account, then, comparing (1.88) and (3.139), we see that the sole quantity that changes is k_z', which is the sum of the nonlinear part (3.141) and the linear part (1.90):

$$k_z'^2 = \left[1 + 4\pi N\left(\chi^{(1)} + \chi^{(3)}E_\omega^2\right)\right]k^2.$$ (3.149)

Since $\Delta k_z \ll k$ remains valid in a rarefied gas [Eq. (1.35)] with a weak nonlinearity [Eq. (3.143)], a modification of (3.149) in no way changes the above conclusions concerning refraction, which is determined by the ratio k_y/k.

3.5.4. Stationary Self-Focusing of Light

Let us turn to the expression (3.130) for the field strength $\mathbf{E}$ in the incident wave. If we modify it slightly, writing it in the form

$$\mathbf{E}(y, z, t) = \mathbf{E}_\omega\left(1 - \frac{|y|}{a}\right)\exp(i\omega t - ikz),$$ (3.150)

with $|y| \le a$, the result describes an incident light beam whose diameter is $2a$. Since there is no dependence on x, this is still a model. The real case of axial symmetry leads to no conceptually new qualitative elements in the conclusion we will arrive at below.

We will take up the case $\chi^{(3)} > 0$, where $|y| < y_0$, that is, the refraction of rays occurs in the direction of the beam's axis $y = 0$ (focusing). According to (3.148), each ray in the beam corresponds to the fixed value of $|y_0| \ll a$, refracting to the beam's axis $y = 0$. Thus, stationary self-focusing of the light beam in the medium emerges as a result of the nonlinearity of the medium (for $\chi^{(3)} < 0$ it is defocusing that emerges).

In view of the symmetry of the beam, we can restrict our discussion to the case of positive y, and it is for this case that (3.148) was written. Putting $y = 0$, we can find the value $z = d$ at which a light ray intersects the beam's axis. Using the language of geometrical optics, we can call d the *focal length*. From the equation for the trajectory of the ray, (3.148), we find a formula for d:

$$d = \left(\frac{ay_0}{2\pi N \chi^{(3)} E_\omega^2} \right)^{1/2} \gg y_0. \qquad (3.151)$$

We see that d depends on the distance y_0 of the ray from the beam's axis $y = 0$, that is, there is aberration of the light beam.

Let us estimate d from above, putting $y_0 \sim a$. Note that this, strictly speaking, is not a correct assumption since (3.151) was obtained on the assumption that $y_0 \ll a$. But the weak dependence of d on y_0 makes it possible to obtain an upper bound on d, at least in order of magnitude. The result is

$$d \sim \left(2\pi N \chi^{(3)} E_\omega^2 \right)^{-1/2} a \gg a. \qquad (3.152)$$

The quantity d is also called the *self-focusing length* (Kelley, 1965). It is one of the main characteristics of self-focusing and is widely used in comparing experimental data with theoretical calculations (Akhmanov, Khokhlov, and Sukhorukov, 1972; Akhmanov, Sukhorukov, and Khokhlov, 1968; Lugovoï and Prokhorov, 1974; Marburger, 1975; Shen, 1975a, b). Note that although the self-focusing process itself is caused by nonlinear refraction and hence is determined by high field strengths in the wave, the self-focusing length depends only weakly on the field strength.

Substituting d from (3.151) for z from (3.146), we arrive at an estimate for the angle α at the focus:

$$\alpha \approx 4\pi \frac{d}{a} N \chi^{(3)} E_\omega^2 = 2 \left[2\pi \frac{y_0}{a} N \chi^{(3)} E_\omega^2 \right]^{1/2}. \qquad (3.153)$$

If $y_0 \sim a$, then from (3.153) and (3.143) it follows that α remains small ($\alpha \ll 1$). Equation (3.153) also shows that although the inequality $y_0 \ll a$ is violated [this inequality was used in deriving the formula (3.152) for the self-focusing length], the substitution of d into (3.146) yields a correct condition imposed on the angle α at the focus, namely, $\alpha \ll 1$. This inequality is equivalent to the initial relationship (3.143), which reflects the approach employed here, the unperturbed-field approximation with respect to the incident wave.

Another often used characteristic of self-focusing is the *critical field strength* (or the *critical radiation power*). Because of the model employed, this quantity did not appear in the above discussion. The incident beam of light was assumed to be strictly parallel. As already noted in Section 3.5.1, any real wave incident on a nonlinear medium is divergent. In real laser radiation the minimal divergence is determined by diffraction (this is known as diffraction-limited divergence). If now we take into account that the incident wave is divergent, we arrive at the concept of the critical field strength, which is the field strength in the incident wave for which self-focusing exceeds diffraction-limited divergence. The angle of diffraction-limited divergence is estimated at $\alpha_d \approx \lambda/a = (ka)^{-1}$, with λ the radiation wavelength, while the self-focusing angle is estimated via (3.153). The condition $\alpha = \alpha_d$ yields the following expression for the critical field strength:

$$E_\omega^{cr} \approx \left(\frac{a}{y_0}\right)^{1/2} (ka)^{-1} \left(8\pi N\chi^{(3)}\right)^{-1/2}. \tag{3.154}$$

Since $ka \gg 1$, Eq. (3.154) does not contradict the condition (3.143) for weak nonlinearity. This formula at $y_0 \sim a$ yields $E_\omega^{cr} \sim (ka)^{-1}(N\chi^{(3)})^{-1/2}$. On the other hand, let us turn to the criterion (3.145), which ensures that the solution (3.140) is valid. Substituting into (3.145) the solution (3.148) for y, the expression for k_y from (3.142), $y \sim y_0$, $z \sim d$, and d from (3.151), we obtain

$$(ka)^2 N\chi^{(3)}E_\omega^2 \ll \left(\frac{a}{y_0}\right)^3, \tag{3.155}$$

which shows that the field strength E_ω is bounded above. The expression (3.154) leads to an inequality bounding the field strength from below:

$$(ka)^2 N\chi^{(3)}E_\omega^2 \gg \frac{a}{y_0}, \tag{3.156}$$

which does not contradict (3.155), because $a \gg y_0$. Neither (3.155) nor (3.156) contradicts the condition (3.143) for weak nonlinearity.

From (3.154) we find the critical radiation power W_{cr} (Ciao, Garmiere, and Townes, 1964) by multiplying radiation energy [proportional to $(E_\omega^{cr})^2$] per unit volume by the speed of light (obtaining the magnitude of the Poynting vector) and then by the beam's cross-sectional area (proportional to a^2). At $y_0 \sim a$ we have

$$W_{cr} \sim \frac{c\lambda^2}{N\chi^{(3)}}. \tag{3.157}$$

Note that the size a of the incident beam is absent from the expression for the critical radiation power. From (3.157) it also follows, as expected, that the greater the nonlinear susceptibility $\chi^{(3)}$, the lower the critical radiation power.

In accordance with the condition (3.156), self-focusing requires that the field strength in the incident wave exceed the critical value (3.154); in this event the focusing angle α exceeds the diffraction angle α_d of the light wave. In the opposite limiting case, $\alpha \ll \alpha_d$, there is obviously no self-focusing; instead there is diffraction-limited divergence. In the intermediate case, where the field strength E_ω is of the order of the critical field strength (3.154), the beam's radius remains practically unchanged in the medium. In this case one speaks of *self-channeling* of the light beam in the medium.

Above we considered the case where $\chi^{(3)}$ is positive. Obviously, if $\chi^{(3)}$ is negative, the phenomenon of nonlinear refraction will manifest itself in the self-defocusing of the incident beam.

The work of Bjorkholm and Ashkin, 1974, may serve as an example of an experiment in which stationary nonlinear refraction was observed in an atomic gas. The radiation from a dye laser with variable frequency was directed into a cell containing sodium vapor. The frequency of the laser radiation was changed near the green doublet in the sodium spectrum (the transition from the ground state $3S_{1/2}$ to the excited state $3P_{3/2}$), so that there was a one-photon resonance, and the nonlinear atomic susceptibility of sodium changed rapidly. At a certain frequency of the incident radiation (in relation to the frequency of the above-noted transition), both self-focusing and self-defocusing of the beam were observed, depending on the misfit.

The variation in nonlinear susceptibility may be caused also by the presence of multiphoton resonances; an example of self-focusing observed in a two-photon resonance has been given by Bakhramov, Gulyamov, Drabovich, and Faĭzulaev, 1975. In their study potassium vapor was irradiated by a dye laser, and the two-photon resonance was observed on the transitions from the ground state $4S_{1/2}$ to states $4D_{5/2}$ and $6S_{1/2}$.

3.5.5. The Instability of Self-Focusing of Light

In the model of self-focusing just described we assumed that the incident wavefront is almost plane. Indeed, the distribution of the field strength over the wavefront is nearly uniform—the amplitude $(1 - y/a)\mathbf{E}_\omega$ differs but little from $\mathbf{E}_\omega$ in view of the fact that $y \ll a$. Since all the quantities we are dealing with are real-valued, the phase of the inhomogeneous part $\mathbf{E}_\omega y/a$ coincides with the phase of the homogeneous part $\mathbf{E}_\omega$. Of course, all these arguments do not contradict the fact that the total field is also real-valued (it is obtained by taking the real part of the complex-valued electric field strength $\mathbf{E}$). In fact, the oscillatory dependence of $\mathbf{E}$ on y in the form of $\exp(iy/a)$ leads to the abovementioned phase $\pi/2$ for $y \ll a$.

The above description of nonlinear refraction can easily be modified to incorporate a possible shift in the phase between the inhomogeneous and homogeneous parts in the incident wave. When this shift is reduced to $\pi/2$, one must simply replace α with $i\alpha$ (or $-i\alpha$) in the above formulas. The solution (3.140) will change drastically as a result. According to (3.142), the wave number k_y along the y axis becomes purely imaginary, with the result that, according to (3.140), the electric field strength E' in the medium grows exponentially fast. Mathematically this means that the self-focusing of the beam is unstable: a small inhomogeneity in the transverse direction begins to grow exponentially after the beam enters the medium. From (3.140) it is clear that this increase becomes noticeable when $k_y y$ is greater than unity, but then it slows down, since the perturbation reaches the value of E_ω and (3.145) ceases to be valid. Using the expression (3.142) for k_y, we can find the distances z_0 (assuming that $y \sim y_0$) at which the instability develops:

$$z_0 \sim \frac{a}{y_0}\left(kN\chi^{(3)}E_\omega^2\right)^{-1}. \tag{3.158}$$

For values of y_0 that are not very small in comparison with α we have

$$z_0 \sim \left(kN\chi^{(3)}E_\omega^2\right)^{-1}. \tag{3.159}$$

Thus, self-focusing is destroyed when the inhomogeneous part of the incident wave is shifted in relation to the homogeneous part. At values of the phase differing from $\pi/2$ the rate of increase of instability is numerically different, of course, but in order of magnitude it is the same.

For low values of a, corresponding to strong inhomogeneity in the transverse direction y, the exponentially increasing instability of the wave disappears, but so does the self-focusing effect (Bespalov and Talanov, 1966). The term with $\partial^2/\partial y^2$ in Maxwell's equation (3.132) becomes predominant, since differentiating twice with respect to y results in a large factor a^{-2}. On the other hand, in Eqs. (3.132) and (3.134) we can neglect the right-hand sides with the nonlinear polarization. Then Eq. (3.134) for the electric field strength in the weak polarization wave yields

$$\left(\frac{\partial^2}{\partial y^2} + \frac{\partial^2}{\partial z^2} - 2ik\frac{\partial}{\partial z} \right) \delta E_\omega(y, z) = 0. \qquad (3.160)$$

Suppose that the dependence of δE_ω on y prior to entrance into the medium is determined by the factor $\exp(iy/a)$ (here we have taken the phase to be $\pi/2$ in relation to the amplitude E_ω of the homogeneous part of the incident wave). Equation (3.160) then shows that this dependence is retained after the beam has entered the medium. We may therefore write $\delta E_\omega(y, z) = \xi(z)\exp(iy/a)$, and (3.160) yields the following equation for $\xi(z)$:

$$\xi(z) = -2ika^2\frac{d\xi(z)}{dz}. \qquad (3.161)$$

Here we have ignored the term with d^2/dz^2, which is small compared to the term with $-2ik(d/dz)$, since their ratio is of the order of $(kz)^{-1} \ll 1$. This corresponds to the approximation of truncated Maxwell equations, of which we spoke earlier.

Equation (3.161) yields the solution

$$\xi(z) \sim \exp(iz/2ka^2). \qquad (3.162)$$

Thus, the weak polarization wave δE_ω retains its oscillatory nature as it enters the medium, with characteristic variations in δE_ω along the z axis occurring on the scale of approximately ka^2 (Bespalov and Talanov, 1966). Hence, in the case of small a, instability is absent but so is self-focusing, as the solution (3.162) shows. When a is small and the phase is zero instead of $\pi/2$, the structure of the solution (3.162) undergoes no qualitative change: a is replaced with ia, and a^2 is replaced with $-a^2$, so that the expression within the parentheses in (3.162) simply changes sign.

For the solution (3.162) to be valid, that is, for a to be small, the range of values of ka^2 corresponding to characteristic values of z in the solution (3.162) must be small compared to the length z_0 in (3.158) over which the instability develops:

$$(ka)^2 N\chi^{(3)}E_\omega^2 \ll \frac{a}{y_0}. \qquad (3.163)$$

We see that the inequality (3.163) is opposite to (3.156).

3.5.6. The Elementary Processes That Lead to the Dependence of the Index of Refraction on the Field Strength in the Light Wave

The above discussion implies that the reason for nonlinear refraction is the nonlinearity in the susceptibility of the medium at the frequency of the incident light wave. Since for the medium we take an atomic gas, we must consider only the odd powers in the expansion of the susceptibility in powers of the field strength. The assumption was that the nonlinearity is introduced by the first (cubic) term and is described by the Feynman diagram depicted in Figure 2.9. In agreement with this diagram, the refractive index is a linear function of the light intensity.

The nonlinear polarization of an atomic gas is a rapidly varying function of the radiation frequency. The resonance maxima in the nonlinear susceptibility appear as a result of resonances between the radiation frequency and the frequency of electron transitions in the spectrum of bound electronic states of an atom or (in the case of a strong field) in the quantum system consisting of an atom and the external field. When a resonance manifests itself, the susceptibility is determined by the population of excited electronic states. The formulas for the nonlinear polarization given above correspond to the case where there is no population of the resonance state. In the case of a resonance we must start with the linear polarization $P_\omega^{(1)}$ of an atomic gas at the frequency of the incident wave, ω. This is given by the following formula:

$$\mathbf{P}_\omega^{(1)} = 2\pi \left[N_n \chi_{nn}^{(1)} + N_m \chi_{mm}^{(1)} \right] \mathbf{E}_\omega, \qquad (3.164)$$

where the label n stands for the ground state of the atom, and m for an excited state. The numbers of atoms in these states, N_n and N_m, are determined by the field strength $\mathbf{E}_\omega$ in the wave; only the total number of all the atoms in the system, $N_n + N_m = N$, is independent of $\mathbf{E}_\omega$. Since $\chi_{nn}^{(1)}$ differs from $\chi_{mm}^{(1)}$, the polarization $\mathbf{P}_\omega^{(1)}$, as shown by Eq. (3.164), is actually a nonlinear function of $\mathbf{E}_\omega$.

The excited state m becomes populated by various one-photon and multiphoton processes. The one-photon processes can be studied within the framework of Eq. (3.164), for instance, within the two-level model (Armstrong, Bloembergen, Ducing, and Pershan, 1962). The effect that level population in two-photon processes has on $\chi^{(3)}$ has been studied in Section 2.2.2. One-photon population of an excited state may be due both to Rayleigh scattering of light falling on the atomic medium (Javan and Kelley, 1966) and Raman scattering (Butylkin, Kaplan, and Khronopulo,

1971). The term used in the first case is *resonance self-focusing*, while in the second it is stimulated-Raman-scattering self-focusing. The difference between the two lies in the limiting population of state m, which in resonance self-focusing cannot exceed $N_m^{max} = N/2$ (the saturation effect). Stimulated hyper-Raman scattering of light (Section 3.3.11) can be cited as a particular example of multiphoton population (Skupsky and Osborn, 1975).

As in all other cases, in describing resonance processes that take place in a strong external field one must take into account the variation in the position of atomic energy levels caused by the dynamic polarizability of the atom. Even in one-photon excitation it is important to allow for dynamic polarizability (Butylkin, Kaplan, and Khronopulo, 1971), and this is especially true in multiphoton excitation, since here the dynamical process constitutes a lower-order process in the number of absorbed photons. A detailed description of processes that lead to dependence of the refractive index on the field strength in the wave caused by the variation in population of the excited states has been carried out by Butylkin, Kaplan, Khronopulo, and Yakubovich, 1984.

In a real situation the width of a resonance is determined not so only by the natural width of the resonance transition but also by the width of the nonmonochromatic-laser-radiation spectrum (Section 1.2.4). As for the natural width of the resonance transition, at high field strengths in the external field it is determined not by the spontaneous transitions but by the induced electron transitions from the resonance state and hence depends on the field strength.

Another possibility for the emergence of nonlinear polarization is the ionization of the atomic gas in the field of the light wave and in free electrons appearing in the gas. The polarization of the gas caused by the appearance of free electrons is described by a relationship that follows from (1.15):

$$\mathbf{P}_e = \frac{e^2 N_e}{m\omega^2}\mathbf{E}_\omega, \tag{3.165}$$

where the free electron concentration N_e is a function of $\mathbf{E}_\omega$. Depending on the type of process that results in the ionization, one-photon ($K = 1$) or multiphoton ($K > 1$), the electron concentration $N_e \propto \mathbf{E}_\omega^{2K}$ and hence the polarization $\mathbf{P}_e \propto \mathbf{E}_\omega^{2K+1}$ depend on $\mathbf{E}_\omega$ via a power law.

In addition to the above reasons for self-focusing of light in a gaseous medium, we will also consider the self-focusing of a light beam caused by the heating of the gas by the incident electromagnetic wave. The dielectric constant ϵ of the gas varies under a small variation in temperature by T' in relation to the unperturbed temperature of the gas, T_0, according to the following obvious law:

$$\epsilon = \epsilon_0 + \frac{d\epsilon}{dT}T', \qquad (3.166)$$

where the derivative is taken at $T = T_0$, and $\epsilon_0 = \epsilon(T = T_0)$.

We will demonstrate that focusing occurs in media with $d\epsilon/dT$ positive, and defocusing otherwise. We start by calculating T'. Since the specific heat of an atomic gas is low, we can ignore it over time intervals of the order of the pulse duration. The thermal energy per unit volume of gas is $c_p\rho T'$, with ρ the gas density and c_p the specific heat at constant pressure. By how much does the energy change per unit time, that is, what is the value of $d(c_p\rho T')/dt$? The change is caused by absorption of energy from the passing light wave. The energy flux density in the wave (the intensity of radiation I_ω) is equal to the magnitude of the Poynting vector, or $I_\omega = c|\mathbf{E}_\omega|^2/2\pi$. Multiplying this quantity by the linear absorption coefficient κ of light in the gas, we obtain the energy κI_ω liberated in a unit volume of gas in one second. Thus,

$$c_p\rho\frac{dT'}{dt} = \frac{\kappa c|\mathbf{E}_\omega|^2}{2\pi},$$

from which we find that T' grows linearly with time:

$$T'(t) = \frac{\kappa c|\mathbf{E}_\omega|^2 t}{2\pi\rho c_p}. \qquad (3.167)$$

If we write the dielectric constant (3.166) in the form $\epsilon = \epsilon_0 + \epsilon_2|\mathbf{E}_\omega|^2$, then from (3.167) we obtain an expression for ϵ_2:

$$\epsilon_2 = \frac{d\epsilon}{dT}\frac{\kappa ct}{2\pi c_p}. \qquad (3.168)$$

Here ϵ_2 plays the role of $4\pi N\chi'^{(3)}$ [e.g. see (3.100)], and t is the pulse duration.

An important feature of thermal effects is that the perturbation of the dielectric constant is proportional to the length of a laser pulse, t (Litvak, 1966), while $N\chi'^{(3)}$ is in no way dependent on time and the self-focusing is determined only by the radiation power of the light beam, (3.156).

Since for $d\epsilon/dT$ positive the expression in (3.168) is real and positive, the dielectric constant ϵ in the beam diminishes as one goes from the beam axis to the periphery, because $|\mathbf{E}_\omega|^2$ diminishes; it therefore has a focusing effect, as a convergent lens has, impeding the development of instability in the wave, a process considered in Section 3.5.5. Hence, thermal effects should mitigate the spontaneous instability in the laser beam (Raĭzer, 1966).

On the other hand, heating results in thermal expansion. Hence, the number of atoms N on the beam axis will be lower than on the periphery (Raĭzer, 1967) because the central region is heated more than the periphery. This effect results in ϵ being lower on the beam axis than on the periphery; in other words, thermal expansion has a defocusing effect. Thermal self-defocusing in a number of cases may become greater than thermal self-focusing (Smith, 1969).

3.5.7. Nonlinear Refraction in the Nonstationary Interaction of Light with the Medium

In line with the general framework of our discussion we will not give a detailed analysis of the case where nonlinear refraction appears in the process of propagation of pulsed laser radiation. From the practical standpoint, however, this is an extremely important case, since it is with pulsed radiation that we can attain radiation powers necessary for observing nonlinear refraction. We will therefore note briefly two important facts that determine the development of nonlinear refraction under pulsed radiation.

First, the pulsed mode has a strong effect on the microscopic reasons for the development of nonlinear polarization of the medium. For instance, simple estimates have shown that already with nanosecond pulses the medium has not enough time to heat up for ϵ to change considerably [see Eq. (3.168)], and the more so for thermal expansion, which propagates with the speed of sound. As for inertialess electronic mechanisms, there are also special features in such processes. For instance, with picosecond pulses the resonance nature of processes caused by electron transitions proves to be inhibited because of the large spectral width of the laser radiation, $\Delta\omega \sim t^{-1}$. With nanosecond pulses of single-mode radiation, when the resonance effects are not suppressed because of the finite spectral width of the radiation, the medium, on the contrary, responds adiabatically to a wave with the field strength $\mathbf{E}_\omega(t)$. The adiabatic response occurs when the frequency of the wave is close to that of an atomic transition between discrete states. In this event in the expression (1.22) for the resonance susceptibility $\chi_{nn}^{(1)}$ we must replace $\omega_{mn} - \omega$ with $\omega_{mn} - \omega + i\Gamma_m$, where Γ_m is the width of state m. In the denominator of the expression for the population of state n there appears the Rabi frequency (see Delone and Kraĭnov, 1985, Section 3.1)

$$\Omega(t) = \left[(\omega_{mn} - \omega)^2 + z_{mn} E_\omega(t)^2 \right]^{1/2}. \tag{3.169}$$

Then in the resonance approximation we have

$$N_n = \frac{1}{2} \left[1 + \left(\frac{\omega_{mn} - \omega}{\Omega} \right)^2 \right]. \tag{3.170}$$

The dependence of N_n on E_ω leads to a nonlinearity and consequently to possible effects of focusing and defocusing in the light beam. The population of state n given by (3.170) changes adiabatically when a laser pulse passes through the medium. Indeed, at first $E_\omega(t)$ increases (the leading edge of the pulse) and the Rabi frequency $\Omega(t)$ becomes effectively higher and N_n smaller, but then $E_\omega(t)$ decreases (the trailing edge of the pulse) and N_n grows. The variation with time of the nonlinear part of the refractive index,

$$n_\omega(E_\omega) = 1 + 2\pi \left(N_n \chi_{nn}^{(1)} + N_m \chi_{mm}^{(1)} \right), \tag{3.171}$$

causes the nonlinear refraction to be nonstationary (Grischkowsky, 1970; Grischkowsky and Armstrong, 1972). Besides, in (3.171) we must allow for a similar variation with time in N_m, which complicates the dependence of n_ω on t. In the resonance approximation, $\chi_{nn}^{(1)} \approx -\chi_{mm}^{(1)}$ if we allow for (1.22). Hence, according to (3.171), the value of n_ω is determined by the difference in populations, $N_n - N_m$.

The second feature that emerges when pulsed laser radiation acts on a medium follows from (3.152). Since d is of the order of $[E_\omega(t)]^{-1}$, after the amplitude of the pulse at the leading edge reaches $E_\omega = E_\omega^{cr}$ defined in (3.154) and focusing sets in, the focus moves in opposition to the laser beam, since E_ω increases and so d decreases.

When the maximum in the temporal distribution in the radiation is reached, the focus will lie closest to the boundary of the medium. As the pulse amplitude at the trailing edge falls off, the focus will move away from the medium boundary as long as the electric field strength E_ω is no lower than the critical value of E_ω^{cr}, at which value self-focusing ceases. Since in reality the focus moves at an extremely high speed, of the order of the ratio of the self-focusing length d to the pulse length t ($\sim 10^9$ cm/s), there is no method of stationary observation of the self-focusing effect by which focus movement can be detected, either in the plane of incidence of radiation on the medium or in a perpendicular plane. Focus movement takes place along a narrow channel lying on the beam axis, so that to observe it one needs experimental techniques with extremely high temporal resolution. Such experiments have substantiated theoretical predictions, which, however, are much broader than the qualitative reasoning presented above. Focus movement has been discussed in detail by Lugovoĭ and Prokhorov, 1974, and Shen, 1975a, b.

3.5.8. Self-Focusing of Laser Radiation in Other Media

There can be no doubt that qualitatively the same phenomena that occur in an atomic gas occur in other media. Of course in molecular gases, liquids, and transparent insulators, the microscopic phenomena leading to nonlinear susceptibility have a different nature; some of these were discussed at the end of Chapter 2. However, the presence of nonlinear susceptibility always opens up the possibility for nonlinear refraction to occur, provided that certain conditions are met, conditions that depend on concrete parameters of medium and radiation. As for self-focusing, this process has been studied most thoroughly in liquids and transparent insulators. A number of liquids constitute convenient media for observing the self-focusing of laser radiation because of their extremely high nonlinear susceptibility. Nitrobenzene is one example.

In examining the self-focusing effect one must bear in mind that there is a considerable increase in the field strength in the light field at the focus compared with the values in the incident light wave. This leads to such effects as nonlinear ionization of the medium and breakdown at the focus. The resulting plasma makes the focus opaque for electromagnetic radiation.

But for ordinary transparent insulators (crystals and glasses) self-focusing is extremely important from the practical viewpoint, since it limits the radiation power that can be created or transmitted through active or structural elements of lasers and devices generating or using laser radiation. For instance, it is self-focusing that determines the radiation resistance of activated glasses used as active elements in high-power solid-state lasers.

In conclusion we note that in considering the main topic discussed in this book, nonlinear optics, one must always bear in mind that nonlinear refraction considerably complicates the design and interpretation of experiments, since it changes the spatial distribution of laser radiation.

3.5.9. Conclusion

If the case of normal incidence of a strong wave on the boundary of the medium is compared with that of a weak wave, we see that in both cases the wave propagating within the medium constitutes a linear combination of the incident wave and the medium-polarization wave. Linear polarization of the medium, which occurs in both cases, changes only the component of the wave vector along the direction of propagation of the wave, while the appearance of a component normal to this direction is due to the nonlinear polarization of the medium. For this reason, refraction appears only in the case of a strong wave and is a result of the high field strength in the light wave created in the medium.

The very fact that nonlinear refraction appears at high field strengths in the incident wave is noteworthy, since it means that the basic law of ordinary linear optics of weak light waves, the law of rectilinear propagation of light, is violated. This law is valid only for weak light waves (low light intensities), when only linear polarization of the medium is important and nonlinear polarization can be neglected.

3.6. CHANGE OF POLARIZATION OF AN INTENSE LIGHT WAVE

3.6.1. Introduction

When a weak light wave with a field strength E propagates in a medium, there emerges a linear polarization $P^{(1)}$ in the medium, with the total field $E + 4\pi P^{(1)}$ having the same polarization as the field E in the incident wave, irrespective of the type of polarization of the incident wave (Section 1.6). But when the wave is intense, the situation changes. The intense wave generates nonlinear polarization in the medium. Generally, as we will subsequently show, the polarization of the total field differs from that of the incident field.

To illustrate this statement qualitatively, we turn to a medium in which, under the action of a strong incident wave with elliptic polarization and field strength E, there emerges a wave caused by cubic polarization $P^{(3)}$ at the frequency of the incident wave. The corresponding nonlinear susceptibility $\chi^{(3)}_{ijkl}(\omega; \omega, -\omega, \omega)$ is described by various Feynman diagrams, one of which is shown in Figure 2.9.

By analogy with the case of linear susceptibility (Section 1.1.3), we can say that the tensor $\chi^{(3)}_{ijkl}$, after being averaged over the various directions of the angular momentum of an atom freely oriented in space, can be expressed only by combining products consisting of pairs of Kronecker deltas with different subscripts:

$$\chi^{(3)}_{ijkl} = a\,\delta_{ij}\,\delta_{kl} + b\,\delta_{ik}\,\delta_{jl} + c\,\delta_{il}\,\delta_{kj}. \tag{3.172}$$

Thus, tensor $\chi^{(3)}_{ijkl}$ contains three independent components, a, b, and c. As in the linear case, this fact follows from the absence of a preferred direction in space.

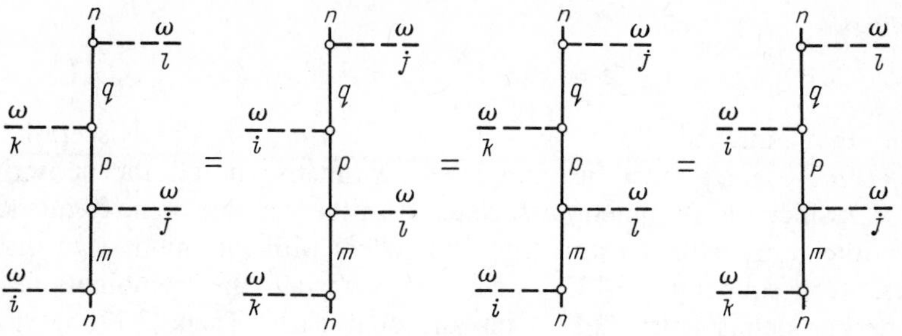

Figure 3.35. Symmetry of the cubic susceptibility $\chi_{ijkl}^{(3)}(\omega; \omega, -\omega, \omega)$ under permutations of i, j, k, and l.

The relationship (3.172) is valid for any tensor $\chi_{ijkl}^{(3)}$, say, for $\chi_{ijkl}^{(3)}(3\omega; \omega, \omega, \omega)$, which corresponds to third-harmonic excitation. The tensor we are interested in here, $\chi_{ijkl}^{(3)}(\omega; \omega, -\omega, \omega)$, however, possesses additional symmetry with respect to permutations $ik \to ki$ and $jl \to lj$. This is illustrated by Figure 3.35, which provides the respective Feynman diagrams: each Feynman diagram describes the absorption of two photons with the ith and kth components of the field strength and the emission of two photons with the jth and lth components. These projections (components) are also shown in Figure 3.35.

The combination of terms in (3.172) that satisfies this symmetry property has the form

$$\chi_{ijkl}^{(3)} = a' \delta_{ik} \delta_{jl} + b'(\delta_{ij} \delta_{kl} + \delta_{il} \delta_{kj}), \qquad (3.173)$$

which in contrast to (3.172) contains only two independent quantities, a' and b'.

Now let us substitute (3.173) into the formula for the cubic polarization corresponding to the Feynman diagram in Figure 2.9 [use Eq. (2.9) with $\omega_1 = -\omega_2 = \omega_3 = \omega$]:

$$P_i^{(3)} = \sum_{jkl} \chi_{ijkl}^{(3)} E_j^* E_k E_l. \qquad (3.174)$$

In accordance with (1.92), the fields E_k and E_l correspond to the absorbed photons and field E_j to the emitted photon. For the cubic polarization vector we arrive at the following formula:

$$P_i^{(3)} = a'|\mathbf{E}|^2 E_i + b'\{\mathbf{E}^2 E_i^* + |\mathbf{E}|^2 E_i\}, \qquad (3.175)$$

or

$$\mathbf{P}^{(3)} = A|\mathbf{E}|^2\mathbf{E} + B\mathbf{E}^2\mathbf{E}^*, \qquad (3.176a)$$

where

$$A = a' + b', \qquad B = b'. \qquad (3.176b)$$

Note that in the limit $\omega \to 0$, the tensor of dc susceptibility $\chi_{ijkl}^{(3)}(\omega; \omega, -\omega, \omega)$ must be completely symmetric in all the subscripts i, j, k, l, since the asymmetry associated with the fact that some frequencies enter into $\chi_{ijkl}^{(3)}$ with the plus sign and others with the minus sign disappears. Consequently, (3.173) yields $a' = b'$; all the combinations of Kronecker deltas enter (3.173) on an equal basis. Then (3.176b) yields $A(\omega = 0) = 2B(\omega = 0)$. The same is approximately true for frequencies that are low compared to the characteristic frequencies of atomic transitions.

Equation (3.176) shows that the direction of the vector $\mathbf{P}^{(3)}$ does not coincide with that of $\mathbf{E}$ in the incident electromagnetic wave, since in the general case of an elliptically polarized wave vector, $\mathbf{E}^*$ is not parallel to $\mathbf{E}$ (Section 1.6.2). Thus, the direction of the electric field strength $\delta \mathbf{E} = 4\pi \mathbf{P}^{(3)}$ in the nonlinear polarization wave does not coincide with that of the electric field strength $\mathbf{E}$ in the incident wave. In determining the electric field strength $\mathbf{E}' = \mathbf{E} + \delta \mathbf{E}$ in the total wave of frequency ω in the medium, we find that the nonlinear medium changes the polarization of the wave. Our aim is to describe this change. Just as before, we will carry out our investigation using the unperturbed-field approximation.

3.6.2. An Equation for the Nonlinear Polarization Wave

Let us write a formula for the electric field strength $\mathbf{E}'$ in an elliptically polarized wave propagating in a medium similar to the formula (1.92) for the electric field strength $\mathbf{E}$ in a wave propagating in a vacuum:

$$\mathbf{E}' = \mathbf{E}'_\omega \left[\cos\left(\theta' + \frac{\pi}{4} \right) \exp(i\varphi') \mathbf{e}_+ \right.$$
$$\left. + \cos\left(\theta' - \frac{\pi}{4} \right) \exp(-i\varphi') \mathbf{e}_- \right]$$
$$\times \exp(i\omega t - ik'z). \qquad (3.177)$$

Thus, we have replaced the unperturbed quantities $\mathbf{E}_\omega$, θ, φ, and k with the perturbed quantities $\mathbf{E}'_\omega$, θ', φ', and k', respectively.

In what follows we will assume that the nonlinear susceptibility $\chi^{(3)}$ is reduced to the quantity $\chi_{ijkl}^{(3)}(\omega; \omega, -\omega, \omega)$, which was discussed in Section 3.6.1. This means that the frequency of the light remains unchanged, that is, the energy transfer from the incident electromagnetic wave to waves of other frequencies (e.g. Raman frequencies or higher harmonics) is neglected. Such an approximation is valid if $\chi_{ijkl}^{(3)}(\omega; \omega, -\omega, \omega)$ is large compared to values of $\chi^{(3)}$ at other frequencies. If this approximation is valid, then $\mathbf{E}'_\omega = \mathbf{E}_\omega$, in view of energy conservation [here $\mathbf{E}_\omega$ is given by (1.95)].

Subtracting the unperturbed field (1.92) from the field perturbed by the medium, (3.177), and assuming $\Delta\theta = \theta' - \theta$, $\Delta\varphi = \varphi' - \varphi$, and $\Delta k = k' - k$ to be small in comparison with θ, φ, and k, we arrive at the following expression for the electric field strength $\delta\mathbf{E} = \mathbf{E}' - \mathbf{E}$ in the polarization wave generated in the nonlinear medium:

$$\delta\mathbf{E} = i\,\Delta k\,z\,\mathbf{E} - \left\{ \Delta\theta\,\mathbf{E}_\omega \left[\cos\left(\theta - \frac{\pi}{4}\right)\exp(i\varphi)\,\mathbf{e}_+ \right. \right.$$
$$\left. - \cos\left(\theta + \frac{\pi}{4}\right)\exp(-i\varphi)\,\mathbf{e}_- \right]$$
$$+ i\,\Delta\varphi\,\mathbf{E}_\omega \left[\exp\left(\theta + \frac{\pi}{4}\right)\exp(i\varphi)\,\mathbf{e}_+ \right.$$
$$\left.\left. - \cos\left(\theta - \frac{\pi}{4}\right)\exp(-i\varphi)\,\mathbf{e}_- \right]\right\}$$
$$\times \exp(i\omega t - ikz). \tag{3.178}$$

Our main goal here is to determine the differences $\Delta\theta$, $\Delta\varphi$, and Δk starting from the Maxwell equations (3.4) for the polarization wave at frequency ω propagating in the medium:

$$\nabla^2\delta\mathbf{E} - \left(\frac{n_\omega}{c}\right)^2\frac{\partial^2\,\delta\mathbf{E}}{\partial t^2} = \frac{4\pi N}{c^2}\frac{\partial^2\mathbf{P}^{(3)}}{\partial t^2}, \tag{3.179}$$

where $\mathbf{P}^{(3)}$ stands for the nonlinear part of the cubic polarization at the frequency ω of the incident wave (Figure 2.9), since, as established in Section 1.6, the linear part has no effect on the angles θ and φ and is taken into account in the left-hand side of (3.179). In the unperturbed-field approximation the polarization $\mathbf{P}^{(3)}$ is generated by the unperturbed incident field $\mathbf{E}$.

Since all the quantities in (3.177) depend on time through the exponential $\exp(i\omega t)$, we can rewrite Maxwell's equation (3.179) in the form [e.g. see (3.6)]

$$\nabla^2\delta\mathbf{E} + k^2\,\delta\mathbf{E} = -4\pi k^2 N\mathbf{P}^{(3)}, \tag{3.180}$$

where we have allowed for the fact that $n_\omega \approx 1$, and where $k = \omega/c$. Substituting (3.175) for $\mathbf{P}^{(3)}$ in (3.180), we obtain an equation for $\delta\mathbf{E}$ in the unperturbed-field approximation:

$$\nabla^2\delta\mathbf{E} + k^2\,\delta\mathbf{E} = -4\pi k^2 N\big(A|\mathbf{E}|^2\mathbf{E} + B\mathbf{E}^2\mathbf{E}^*\big). \tag{3.181}$$

3.6.3. Self-Rotation of the Polarization Ellipse of an Intense Light Wave in a Medium

How does one solve Eq. (3.181)? If we express δE in the form similar to (1.92),

$$\delta \mathbf{E} = \delta \mathbf{E}_\omega \exp(i\omega t - ikz), \qquad (3.182)$$

and substitute this into (3.181), we arrive at an equation for $\delta \mathbf{E}_\omega$:

$$\nabla^2 \delta \mathbf{E}_\omega - 2ik\frac{\partial \delta \mathbf{E}_\omega}{\partial z} = -4\pi k^2 N \left[A|\mathbf{E}_\omega|^2 \mathbf{E}_\omega + B\mathbf{E}_\omega^2 \mathbf{E}_\omega^* \right]. \quad (3.183)$$

Note that the complex-valued vector $\mathbf{E}_\omega$ is specified by (1.95).

Next we assume (as we did in other cases) that the nonlinear medium occupies the half space $z > 0$, while the wave with field strength $\mathbf{E}$ falls normally on the interface at $z = 0$ from the left ($z < 0$). The solution of Eq. (3.183) with the boundary condition $\delta \mathbf{E}_\omega(z = 0) = 0$ has the simple form

$$\delta \mathbf{E}_\omega = 2\pi i Nkz \left[A|\mathbf{E}_\omega|^2 \mathbf{E}_\omega + B\mathbf{E}_\omega^2 \mathbf{E}_\omega^* \right]. \qquad (3.184)$$

Substituting (3.178) into the left-hand side and (1.95) into the right-hand side, we can rewrite (3.184) thus:

$$
\begin{aligned}
i\,\Delta k\, z &\left[\cos\left(\theta + \frac{\pi}{4} \right)\exp(i\varphi)\,\mathbf{e}_+ + \cos\left(\theta - \frac{\pi}{4} \right)\exp(-i\varphi)\,\mathbf{e}_- \right] \\
&- \Delta\theta \left[\cos\left(\theta - \frac{\pi}{4} \right)\exp(i\varphi)\,\mathbf{e}_+ - \cos\left(\theta + \frac{\pi}{4} \right)\exp(-i\varphi)\,\mathbf{e}_- \right] \\
&+ i\,\Delta\varphi \left[\cos\left(\theta + \frac{\pi}{4} \right)\exp(i\varphi)\,\mathbf{e}_+ - \cos\left(\theta - \frac{\pi}{4} \right)\exp(-i\varphi)\,\mathbf{e}_- \right] \\
&= 2\pi i NkzE_\omega^2 \Bigg\{ B\cos 2\theta \left[\cos\left(\theta + \frac{\pi}{4} \right)\exp(-i\varphi)\,\mathbf{e}_- \right. \\
&\qquad\qquad\qquad \left. + \cos\left(\theta - \frac{\pi}{4} \right)\exp(i\varphi)\,\mathbf{e}_+ \right] \\
&\qquad\quad + A \left[\cos\left(\theta + \frac{\pi}{4} \right)\exp(i\varphi)\,\mathbf{e}_+ \right. \\
&\qquad\qquad\quad \left. + \cos\left(\theta - \frac{\pi}{4} \right)\exp(-i\varphi)\,\mathbf{e}_- \right] \Bigg\}. \qquad (3.185)
\end{aligned}
$$

If we collect the terms that belong to $\mathbf{e}_+$ and those that belong to $\mathbf{e}_-$, we arrive at two scalar equations:

$$(i\,\Delta kz + i\,\Delta\varphi)\cos\left(\theta + \frac{\pi}{4}\right) - \Delta\theta\cos\left(\theta - \frac{\pi}{4}\right)$$
$$= 2\pi ikzNE_\omega^2\left[B\cos 2\theta\cos\left(\theta - \frac{\pi}{4}\right) + A\cos\left(\theta - \frac{\pi}{4}\right)\right], \quad (3.186)$$

$$(i\,\Delta kz - i\,\Delta\varphi)\cos\left(\theta - \frac{\pi}{4}\right) + \Delta\theta\cos\left(\theta + \frac{\pi}{4}\right)$$
$$= 2\pi ikzNE_\omega^2\left[B\cos 2\theta\cos\left(\theta + \frac{\pi}{4}\right) + A\cos\left(\theta - \frac{\pi}{4}\right)\right]. \quad (3.187)$$

Separating the real and imaginary parts in Eqs. (3.186) and (3.187), we get

$$\Delta\theta\cos\left(\theta - \frac{\pi}{4}\right) = \Delta\theta\cos\left(\theta + \frac{\pi}{4}\right) = 0,$$

which implies that $\Delta\theta = 0$.

Thus, when the wave enters the nonlinear medium, the polarization ellipse retains its shape (the semiaxis ratio $\tan\theta$ does not change). This statement remains valid even if we do not resort to the unperturbed-field approximation.

With (3.188) in mind, we can simplify Eqs. (3.186) and (3.187):

$$\Delta kz + \Delta\varphi = 2nkzNE_\omega^2\left[A + B\cos 2\theta\,\frac{\cos(\theta - \pi/4)}{\cos(\theta + \pi/4)}\right], \quad (3.188)$$

$$\Delta kz - \Delta\varphi = 2\pi kzNE_\omega^2\left[A + B\cos 2\theta\,\frac{\cos(\theta + \pi/4)}{\cos(\theta - \pi/4)}\right]. \quad (3.189)$$

We can now find Δk by adding Eqs. (3.188) and (3.189):

$$\Delta k = 2\pi kNE_\omega^2(A + B). \quad (3.190)$$

We see that Δk and hence the change in the refractive index of the medium do not depend on the ellipticity θ of the field or the angle φ characterizing the orientation of the axes of the polarization ellipse in relation to the Cartesian axes associated with the medium. We also note that Δk is determined by the sum $A + B$, with A and B defined in (3.177).

The quantity $\Delta\varphi$ can be obtained by subtracting (3.189) from (3.188):

$$\Delta\varphi = 2\pi kzNBE_\omega^2\sin 2\theta = \Delta k' z, \quad (3.191)$$

where

$$\Delta k' = 2\pi k N B E_\omega^2 \sin 2\theta. \tag{3.192}$$

Equation (3.191) shows that the orientation of the ellipse axes changes as the wave propagates through the nonlinear medium. This phenomenon is commonly known as *self-rotation of the polarization ellipse* of a strong wave in a nonlinear medium. The angle of self-rotation depends on the path z traveled by the wave in the medium and is also a linear function of the radiation intensity $I_\omega \propto E_\omega^2$. The magnitude of self-rotation is determined by the constant B in the nonlinear polarization (3.176) and is proportional to the density N of the atoms constituting the gas. Finally, the self-rotation effect also depends on the polarization θ of the incident radiation.

For linearly polarized light, $\theta = 0$ and (3.191) yields $\Delta\varphi = 0$, that is, the polarization plane remains fixed, and there is no self-rotation. At $\theta \neq 0$ the direction of self-rotation is determined by the sign of θ. The self-rotation effect is at its maximum for a circularly polarized wave, when $\theta = \pm\pi/4$ and $\sin 2\theta = \pm 1$; however, rotation of a circle cannot be observed in principle.

Thus, if the incident wave is polarized linearly or circularly, the polarization of the wave propagating in a nonlinear medium remains unchanged. In the general case of elliptic polarization, the polarization-ellipse axes undergo a rotation , while the ellipse is not deformed ($\theta = $ const); the angle of rotation $\Delta\varphi$ grows linearly as the coordinate z grows in the direction of propagation of the wave in the medium.

Note that although all the results were obtained on the assumption that there is only cubic polarization $\mathbf{P}_\omega = \mathbf{P}_\omega^{(3)}$ in the medium, they remain valid for a medium with arbitrary nonlinearity. The solution for the general case greatly simplifies if we take a two-level atom and a resonance external field (Arutyunyan, Kanetsyan, and Chaltykyn, 1972; Muradyan, Adonts, and Kolomiets, 1973).

The experiment conducted by Arutyunyan, Papazyan, Adonts, Karmanyan, Ishkhanyan, and Khol'ts, 1975, enabled observing the effect of self-rotation in potassium vapor in conditions of a one-photon resonance with the D lines of the principal doublet in $4S_{1/2} \rightarrow 4P_{1/2,3/2}$.

3.6.4. Induced Optical Anisotropy

The effect of self-rotation of the polarization ellipse accompanying the propagation of an elliptically polarized strong wave in a medium is essentially equivalent to the effect of optical anisotropy of a medium that is isotropic in the absence of a wave. Indeed, let us rewrite the expression

(3.177) for the electric field strength $\mathbf{E}'$ in a nonlinear medium, allowing for $\Delta\theta = 0$ and the expressions (3.190) for Δk and (3.191) for $\Delta\varphi$:

$$\mathbf{E}' = \mathbf{E}_\omega\left[\cos\left(\theta + \frac{\pi}{4}\right)\exp(i\varphi)\,\mathbf{e}_+\exp(i\omega t - ik_+z)\right.$$
$$\left. + \cos\left(\theta - \frac{\pi}{4}\right)\exp(-i\varphi)\,\mathbf{e}_-\exp(i\omega t - ik_-z)\right], \quad (3.193)$$

where we have introduced the notation

$$k_+ = k + \Delta k' + \Delta k, \qquad k_- = k + \Delta k - \Delta k', \quad (3.194)$$

and $\Delta k'$ is defined in (3.192).

Equation (3.193) represents an elliptically polarized wave in the form of a linear combination of two circularly polarized waves: a clockwise-polarized and a counterclockwise-polarized. From (3.194) it follows that these waves have different wave numbers, or different refractive indices:

$$n_+ = 1 + 2\pi N E_\omega^2(A + B + B\sin 2\theta),$$
$$n_- = 1 + 2\pi N E_\omega^2(A + B - B\sin 2\theta). \quad (3.195)$$

[Here, as in the foregoing, we ignored the linear susceptibility of the medium. If we wish to allow for the linear susceptibility $\chi^{(1)}$, we must add $2\pi N\chi^{(1)}$ to the right-hand sides of (3.195).] Thus, an elliptically polarized strong wave (e.g. a circularly polarized wave) induces a nonlinear polarization in the medium that results in optical anisotropy—the refractive index of the medium becomes dependent on the polarization of the transmitted light.

The existence of two refractive indices for two circularly polarized waves with different directions of rotation is known as *induced gyrotropy*. [This term appeared by analogy with common gyrotropy, whereby in a homogeneous gaseous medium placed in a strong constant magnetic field, the permittivity tensor of the medium becomes asymmetric (e.g. see Born and Wolf, 1975).]

Since the coefficients A and B that determine the nonlinear susceptibility and the refractive indices n_+ and n_- have imaginary parts (in addition to real parts), not only do the refractive indices differ, but so do the absorption coefficients. The difference that exists between the absorption coefficients for two circularly polarized waves with different senses of rotation is known as *induced circular dichroism*.

3.6.5. Induced Dichroism and Gyrotropy

To observe optical anisotropy, which appears in a medium under the action of an incident elliptically polarized strong wave, it is advisable to use the second, weak test wave and study its propagation. The polarization of the medium induced by the weak test field can be ignored. Let us see how the nature of propagation of a weak test field changes when a strong wave is incident on the nonlinear medium.

Optical anisotropy created in a medium by an elliptically polarized strong wave changes the polarization of a linearly polarized test wave propagating along the z axis. The test wave becomes elliptically polarized because of induced dichroism, while induced gyrotropy results in rotation of the polarization plane of the test wave. Let us examine how such effects emerge.

We denote the electric field strength in the test wave by $\mathbf{E}_1$, with $E_1 \ll E$. Assuming the wave to be linearly polarized along the x axis, we have

$$\mathbf{E}_1 = E_1 \mathbf{e}_x \exp(i\omega_1 t - ik_1 z), \qquad k_1 = \frac{\omega_1}{c}. \qquad (3.196)$$

Let us see how such a field changes in a medium with refractive indices defined by (3.195). We write the field (3.196) in the form of two clockwise- and counterclockwise-polarized electromagnetic waves with equal amplitudes:

$$\mathbf{E}_1 = 2^{-1/2} E_1 (-\mathbf{e}_+ + \mathbf{e}_-) \exp(i\omega_1 t - ik_1 z). \qquad (3.197)$$

This formula gives the field in the test wave in vacuum. On entering a gaseous medium, in accordance with (3.195), the field changes because of induced optical anisotropy of the medium.

Denoting the test field in the medium by $\mathbf{E}_1'$, we write it in the obvious form

$$\begin{aligned}
\mathbf{E}_1' = 2^{-1/2} E_1 \big[&-\mathbf{e}_+ \exp(i\omega_1 t - in_+ k_1 z) \\
&+ \mathbf{e}_- \exp(i\omega_1 t - in_- k_1 z) \big].
\end{aligned} \qquad (3.198)$$

Substituting (3.195) into (3.198) yields

$$\begin{aligned}
\mathbf{E}_1' = 2^{-1/2} E_1 \exp(&i\omega_1 t - ik_1 z - i\,\Delta k\,z) \\
\times \big[&-\mathbf{e}_+ \exp(-i\,\Delta k'\,z) + \mathbf{e}_- \exp(i\,\Delta k'\,z) \big].
\end{aligned} \qquad (3.199)$$

We see that on entering a nonlinear medium, the linearly polarized test field changes in such a manner that its polarization plane rotates, with the

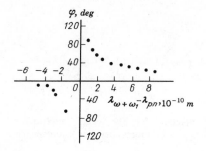

Figure 3.36. Angle of rotation of the linearly polarized test field after passing through a cell containing sodium vapor (Liao and Bjorklund, 1977) as a function of the misfit $\lambda_{\omega+\omega_1} - \lambda_{pn}$, where λ_ω is the wavelength of the incident wave, and λ_{pn} is the wavelength in the $3S \to 5S$ transition.

rotation angle increasing with z, the path traveled by the radiation in the medium, at a rate equal to $\Delta k'$ [see Eq. (3.192)]. The effect is most pronounced when the strong field is circularly polarized, since $\Delta k'$ is at its maximum in this case, according to (3.192). Note that when the circular polarization changes sign, $\Delta k'$ also changes sign, according to (3.192), so that in the process the sense of rotation of the polarization plane of the linearly polarized test field is reversed.

Liao and Bjorklund, 1977, have conducted an experiment in which induced optical anisotropy of a medium was observed by studying the propagation of a weak test wave in this medium. Sodium vapor was irradiated by the radiation of two dye lasers, one of which created the strong field and the other the weak. The frequency of the strong field could be varied near the frequency that corresponds to the transition in the sodium spectrum between the ground state $n = 3S$ and the excited state $p = 5S$. The polarization of the test field was linear. The researchers studied the orientation of the polarization plane of the test field as a function of the frequency (the tuning to the two-photon resonance) and the field strength of the strong wave. Figure 3.36 illustrates the dependence of the angle of rotation of the linearly polarized test field after passing through a cell (of given length) containing sodium vapor on the misfit between the two-photon resonance frequency 2ω and the transition frequency ω_{pn}. As can be seen, the sense of rotation of the polarization plane changes sign at zero misfit, while any increase in the misfit diminishes the effect. This agrees with the theoretical predictions, according to which $\Delta k'$, defined in (3.192), changes sign because the factor B does (the reader will recall that B constitutes a term in the cubic susceptibility). Indeed, for the Feynman diagram in Figure 3.37 describing the resonance term in $\chi^{(3)}(\omega_1; \omega, \omega_1, -\omega)$ we have

$$\chi^{(3)}(\omega_1; \omega, \omega_1, -\omega) \propto [\omega_1 + \omega - \omega_{pn}]^{-1}. \qquad (3.200)$$

The resonance misfit obviously determines the value of the factor B, too.

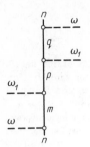

Figure 3.37. The Feynman diagram for the cubic susceptibility $\chi^{(3)}(\omega_1; \omega, \omega_1, -\omega)$ corresponding to the process depicted in Figure 3.36.

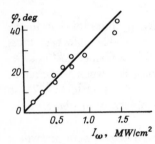

Figure 3.38. Angle of rotation of the linearly polarized test wave after passing through a cell containing sodium vapor (Liao and Bjorklund, 1977) as a function of the intensity I_ω of the strong circularly polarized field with a fixed frequency. The straight line corresponds to theoretical predictions.

Figure 3.38 shows the dependence of the same angle of rotation of the test wave on the intensity of the circularly polarized strong field at a fixed frequency. It can be seen that this dependence is linear, as predicted by (3.192).

In the neighborhood of resonances, the nonlinear susceptibility $\chi^{(3)}$ and the factor in the formula (3.195) for the refractive indices become essentially complex-valued: the imaginary part of the susceptibility corresponds to real absorption in the course of a resonance transition. The same, of course, holds for the coefficient A in (3.195). This formula shows that both n_+ and n_- acquire imaginary parts, having different values. As a result the medium absorbs part of the test wave, and inside the square brackets in (3.199) there appear factors of e_+ and e_- exponentially decreasing as z grows. Thus, the modulus of the first term inside the square brackets becomes different from the modulus of the second term. From the standpoint of the general dependence (3.177) this means that θ is now different from zero, that is, the weak wave becomes elliptically polarized. The ellipticity of the test wave has been calculated by Arutyunyan and Adonts, 1979, who also studied the mechanism of spectral-line broadening in a resonance, a mechanism that leads to absorption. It was found that a one-photon resonance is broadened via the Doppler mechanism, which may be replaced by collision broadening, depending on the temperature of the gas. In a two-photon resonance and with counterpropagating beams of the test wave and the pumping wave with approximately equal frequencies, the linear Doppler effect vanishes (Adonts and Kocharyan, 1976) and only homogeneous (radiation or collision) broadening remains.

3.6.6. Induced Dichroism in Molecular Gases

Induced dichroism may occur not only in atomic gases (the case just discussed), but also in molecular gases. A strong field resonantly spans two vibrational levels in a molecule. Each rotational spectral term contains a system of rotational levels, with the result that the strong field in the incident wave must be tuned in resonance with the difference between two rotational levels belonging to different vibrational terms, while the test field must be tuned in resonance with the difference between one of the above-noted rotational levels and a rotational level belonging to another vibrational term. In addition, there is rotational relaxation, as a result of which the rotational levels within each vibrational term are mixed. Marcano and Garsia-Golding, 1982, have found that the shape of the absorption line of the test wave, considered as a function of the misfit between the rovibronic-molecular-transition frequency and the resonance frequency, instead of Lorentzian becomes double-humped, with the dip exactly in the middle.

3.6.7. Nonlinear Polarization Spectroscopy

Wieman and Hänsch, 1976, have developed a high-resolution spectroscopy method, free from Doppler broadening, that is based on the above-described phenomena of induced dichroism and gyrotropy. The method consists in recording the interaction of two monochromatic beams of laser radiation propagating in a nonlinear absorbing gaseous medium, as a function of the polarization of the light beams.

As noted earlier, the linearly polarized test wave when propagating in the medium becomes elliptically polarized; this is accompanied by rotation of the polarization-ellipse axes. The signal from the test wave that has passed through the medium and then the polarizer results entirely from the anisotropy induced by the strong field and therefore has a nonlinear origin. The influence of Doppler broadening is eliminated if we measure only the nonlinear part of the signal. Note that the method does not require counterpropagating beams (Vasilenko, Chebotaev, and Shishaev, 1970).

Wieman and Hänsch, 1976, used the above method to determine the Stark energy shifts in hydrogen atoms. They considered the spectrum lines representing the two-photon transition $n = 4 \rightarrow n = 2$. The experiment made it possible to measure the Stark shifts of spectral lines in a constant electric field.

Shalagin, 1977, generalized the method of nonlinear polarization spectroscopy to incorporate the case of degenerate atomic levels. This is important because the levels of real atoms and molecules are degenerate in the directions of angular momentum, and the population of sublevels may not be the same. This nonuniformity in population is characterized by "polarization moments": the total population of the level, the orientation, and the alignment (Kastler, 1954). The method of nonlinear polarization spectroscopy is most promising for determining these quantities (Shalagin, 1977).

3.6.8. Conclusion

In Section 1.6 we proved, starting from the assumption of the linear nature of polarization of the wave induced in a medium, that the wave retains its polarization when it travels through a transparent medium (this fact is well known from linear optics). But when the light incident on the medium is strong, nonlinear polarization is induced in the medium (in addition to the linear polarization). If the incident strong wave is linearly polarized, its polarization is not changed by the medium, just as in the case of a weak field. This fact is due to the symmetry of the system consisting of the isotropic medium and the linearly polarized wave. If the incident wave is elliptically polarized, asymmetry emerges. If an elliptically polarized strong wave propagates through the medium, the latter becomes optically anisotropic, with all the consequences that follow from this. For instance, the polarization ellipse of the incident wave rotates as the wave propagates through the medium, and the polarization of the wave varies periodically in the laboratory system of coordinates.

3.7. NONLINEAR LIGHT ABSORPTION

3.7.1. Introduction

Nonlinear absorption is a much more complicated process than linear light absorption, discussed in Section 1.4. The main difference that nonlinearity makes is the dependence of the absorption coefficient on the light intensity, which means that Bouguer's law is invalid for nonlinear absorption. While linear absorption is caused by resonance one-photon excitation of the atom, nonlinear absorption is caused by multiphoton excitation and multiphoton ionization of the atom. Of course, as in the case of one-photon excitation, multiphoton excitation is realized with maximum probability when a multiphoton resonance is present, that is, when $K\omega \approx \omega_{pn}$, with $K = 2, 3, \ldots$. The same is clearly true of multiphoton ionization of atoms, since the ionization probability depends strongly on the frequency when intermediate resonances (including multiphoton resonances) appear. As with every multiphoton process, neither multiphoton excitation nor multiphoton ionization is a threshold process in the electric field strength in an external wave. In principle, any field, however small, may result in a nonzero probability for these processes. But, practically speaking, these processes become important only in strong fields, and—most important—their probability depends on the light intensity in a nonlinear manner (via a power-function law).

Thus, to describe nonlinear light absorption, we must investigate multiphoton excitation and multiphoton ionization of atoms. In this section, however, we will simplify matters by considering only two-photon excitation. For excitation involving a greater number of photons, there is complete qualitative analogy with two-photon excitation. For details on multiphoton ionization we refer the reader to Section 2.4, which contains extensive information on the subject, and to Section 3.2.6, where this process was considered as competing with the excitation of higher optical harmonics.

Thus, we start with two-photon absorption, already discussed as a competitive process in Section 3.2.6 in connection with harmonic excitation. There we found that, as in the case of linear absorption, nonlinear light absorption in a medium is determined by the imaginary part of the susceptibility, χ''. While linear absorption (Section 1.4) is determined by $\chi''^{(1)}(\omega; \omega)$, nonlinear absorption is determined by $\chi''^{(3)}(\omega; \omega, -\omega, \omega)$, which follows, quite obviously, from (1.67) if we substitute $\chi^{(1)} + \chi^{(3)}E_\omega^2$ for $\chi^{(1)}$.

In discussing two-photon absorption as a competing or limiting process in harmonic excitation (Section 3.2), stimulated hyper-Raman scattering (Section 3.3), and nonlinear wave coupling (Section 3.4), we noted that as the resonance $2\omega = \omega_{pn}$ is approached, both the probability of the nonlinear processes discussed in the respective sections and the probability of two-photon absorption increase. For this reason the comparative role of two-photon absorption requires a detailed analysis, which we will now undertake.

3.7.2. The Two-Photon Absorption Probability

The rate of the two-photon transition $n \to p$ can be found if the general formula (2.45) for the probability of multiphoton excitation is employed. In the two-photon case it assumes the form (in the general case of distinct waves)

$$w_{np}^{(2)} = 2\pi \left| \chi_{pn}^{(1)} \right|^2 (E_1' E_2')^2 \rho_p, \tag{3.201}$$

where $\chi_{pn}^{(1)}$ is an off-diagonal element of the linear susceptibility (a two-photon transition matrix element) represented by the first diagram in Figure 2.15, that is,

$$\chi_{np}^{(1)} = \sum_m z_{nm} z_{mp} \left[(\omega_{mn} - \omega_1)^{-1} + (\omega_{mn} - \omega_2)^{-1} \right], \tag{3.202}$$

and ω_1, ω_2, E_1', and E_2' are the frequencies and amplitudes of the fields in the medium.

The formula (3.201) describes both two-photon excitation and two-photon ionization. The differences lie in the energy conservation law and the expression for the density ρ_p of the final states.

For two-photon excitation the energy conservation law has the obvious form $\omega_1 + \omega_2 \approx \omega_{pn}$, and ρ_p in (3.201) is a factor that describes the spectral lineshape of two-photon absorption; according to (2.31) we have

$$\rho_p = \frac{\Gamma_p}{\pi} \left[(\omega_{pn} - \omega_1 - \omega_2)^2 + \Gamma_p^2 \right]^{-1}, \tag{3.203}$$

where, in contrast to one-photon excitation (Section 1.4), the width Γ_p may be conditioned not only by the spontaneous relaxation of state p into the initial state n but also by ionization from the state p (one-photon or multiphoton).

Substituting (3.202) and (3.203) into (3.201), we obtain for the probability of two-photon absorption the following final expression:

$$w_{np}^{(2)} = 2\Gamma_p (E_1' E_2')^2 \left[(\omega_{pn} - \omega_1 - \omega_2)^2 + \Gamma_p^2 \right]^{-1}$$

$$\times \left| \sum_m z_{nm} z_{mp} \left[(\omega_{mn} - \omega_1)^{-1} + (\omega_{mn} - \omega_2)^{-1} \right] \right|^2. \tag{3.204}$$

In (3.204) and (3.202) we have assumed, for the sake of simplicity, that both electromagnetic waves are linearly polarized along the z axis.

In the case of nonresonance two-photon ionization, which is realized when $\omega_1 + \omega_2 > \mathscr{E}_n$, the density ρ_p in (3.201) is determined by the density of the final states of the electron in the continuous spectrum [see Eq. (2.39)].

3.7.3. Saturation of Nonlinear Absorption

Equation (3.204) determines the electron transition rate $w_{np}^{(2)}$ from state n to state p. The absolute probability of the population of state p over time t is therefore determined by the product $w_{np}^{(2)}t$. In Section 3.2.6 we noted [see Eq. (3.48)] that when $w_{np}^{(2)}t$ becomes greater than unity, the populations N_p and N_n of states p and n become closer in value, and in the limit they are both equal to $N/2$. The above inequality determines the *saturation of two-photon transitions* (Section 3.2.6). If we denote by T the time interval during which the laser pulse operates and define t as $\min(T, \Gamma_p^{-1})$, then the condition for the saturation of a two-photon transition takes the form

$$w_{np}^{(2)}t > 1. \tag{3.205}$$

When this is achieved, we have $N_p \approx N_n \approx N/2$, and according to the results obtained in Section 2.2.2, all cubic susceptibilities $\chi^{(3)}$ vanish [e.g. see the expression (2.28) for $\chi^{(3)}(\nu; \omega_1, \omega_2, \omega_3) \propto (N_n - N_p)$]. Thus, the condition (3.205) determines the maximum values of the field strengths, E_1' and E_2', which are optimal for any resonance processes of four-wave coupling (for details see Section 3.7.5):

$$\left(E_1'E_2'\right)_{\max}^2 \sim \frac{\Gamma_p}{t\left|\chi_{np}^{(1)}\right|^2}. \tag{3.206}$$

In this estimate the misfits were restricted to the range $|\omega_1 + \omega_2 - \omega_{pn}| \sim \Gamma_p$ and Eq. (3.204) was used for the transition rate.

3.7.4. The Nonlinear Absorption Coefficient

Let us turn to the general expression for the field strength $\mathbf{E}'$ in the nonlinear polarization wave at the frequency $\nu = \omega_1 + \omega_2 + \omega_3$ that encompasses all the cases of four-wave coupling discussed above:

$$\mathbf{E}'(z, t) = \mathbf{E}_\nu \exp(i\nu t - ik_\nu'z), \tag{3.207}$$

where k_ν' is the wave number of the polarization wave in the medium, defined as $n_\nu \nu/c$. The refractive index n_ν has the form

$$n_\nu = 1 + 2\pi\left[N_n\chi_{nn}^{(1)}(\nu; \nu) + N_p\chi_{pp}^{(1)}(\nu; \nu)\right]. \tag{3.208}$$

This formula was given in Section 3.6 when we discussed the phenomenon of nonlinear refraction. In conditions of two-photon absorption, when

$\omega_{pn} \approx \omega_1 + \omega_2$, the populations N_n and N_p depend on the field strengths E_1' and E_2' in the electromagnetic waves in the medium, with the result that in reality n_ν defined by (3.208) is a nonlinear characteristic and does not coincide with the common linear refractive index

$$n_\nu = 1 + 2\pi N \chi_{nn}^{(1)}(\nu; \nu). \qquad (3.209)$$

Since $\chi^{(1)}$ is real-valued, n_ν is real-valued too, according to (3.208), that is, the refractive index does not lead to absorption, because it was assumed that no one-photon resonances, which are responsible for $\chi^{(1)}(\nu; \nu)$ having an imaginary part, are present (such one-photon resonances were studied in detail in Section 1.4 in connection with linear absorption).

Nonlinear absorption does occur if to (3.208) we add an additional term contributed by the imaginary part of the nonlinear resonance susceptibility $\chi^{(3)}$. To simplify the following consideration, we assume that the wave with field strength E_1' and frequency ω_1 propagating in the medium is strong, while the wave with field strength E_2' and frequency ω_2 is weak; we consider this weak wave as propagating within the nonlinear medium, so that $\nu = \omega_2$ and $\omega_3 = -\omega_1$. All this requires that in (3.208) one must replace ν with ω_2, and

$$\chi_{nn}^{(1)}(\omega_2; \omega_2) \rightarrow \chi_{nn}^{(1)}(\omega_2; \omega_2) + \chi_{nn}^{(3)}(\omega_2; \omega_1, \omega_2, -\omega_1)E_1'^2. \qquad (3.210)$$

The same must be done to $\chi_{pp}^{(1)}$. Substituting (3.210) into (3.208), we find that

$$n(\omega_2) = 1 + 2\pi N_n \left[\chi_{nn}^{(1)}(\omega_2; \omega_2) + \chi_{nn}^{(3)}(\omega_2; \omega_1, \omega_2, -\omega_2)E_1'^2 \right]$$
$$+ 2\pi N_p \left[\chi_{pp}^{(1)}(\omega_2; \omega_2) + \chi_{pp}^{(3)}(\omega_2; \omega_1, \omega_2, -\omega_2)E_1'^2 \right]. \qquad (3.211)$$

When there is a two-photon resonance between states n and p, we have, according to the results obtained in Section 2.2.2, the following:

$$\chi_{nn}^{(3)} = -\chi_{pp}^{(3)} = \chi_{np}^{(1)}(-\omega_2; \omega_1)^2 (\omega_{pn} - \omega_1 - \omega_2 + i\Gamma_p)^{-1}, \qquad (3.212)$$

where the off-diagonal linear susceptibility is

$$\chi_{np}^{(1)}(-\omega_2; \omega_1) = \sum_m z_{nm} z_{mp} \left[(\omega_{mn} - \omega_1)^{-1} + (\omega_{mn} - \omega_2)^{-1} \right]. \qquad (3.213)$$

Next we assume that the distance z traveled by the waves in the medium is small, so that there is no significant energy transfer from the wave with frequency ω_1 to the wave with frequency ω_2. Actually this means that the unperturbed-field approximation in the field with frequency ω_1 is valid here, with the result that the absorption coefficient does not depend on distance z. One must bear in mind, however, that in typical nonlinear-optics problems there is always considerable energy transfer in the course of wave coupling, so that actually the absorption coefficient is a function of z.

Let us return to the case of small energy transfers between the coupled waves. The quantity given by (3.212) has an imaginary part, which leads to nonlinear absorption of the wave with frequency ω_2 in the medium via an exponential law (Section 1.4):

$$I'(\omega_2) = I(\omega_2)\exp(-\kappa z), \tag{3.214}$$

where $I'(\omega_2)$ is the intensity of the wave with frequency ω_2 in the medium, $I(\omega_2)$ is the intensity of this wave before entering the medium, and κ is the absorption coefficient. Combining (3.207) and (3.214) and allowing for the fact that $I' = c|\mathbf{E}'|^2/2\pi$, we find that

$$\kappa = -2\,\text{Im}\,k'(\omega_2) = -\frac{2\omega_2}{c}\,\text{Im}\,n(\omega_2). \tag{3.215}$$

Substituting (3.211) into (3.215) yields the final expression for κ:

$$\kappa = -\frac{4\pi\omega_2}{c}(N_n - N_p)E_1^2\,\text{Im}\,\chi_{nn}^{(3)}(\omega_2;\omega_1,\omega_2,-\omega_1). \tag{3.216}$$

We see that the nonlinear absorption coefficient depends on the field strength E_1 and is proportional to the difference in the numbers N_n and N_p of particles in states n and p per unit volume. When resonance saturation sets in ($N_n \approx N_p \approx N/2$), κ vanishes, which results in nonlinear "clearing up" of the medium.

Taking the imaginary part of $\chi_{nn}^{(3)}$ from (3.212) and substituting it into (3.216) finally yields

$$\kappa = \frac{4\pi\omega_2}{c}(N_n - N_p)E_1^2\left|\chi_{np}^{(1)}(-\omega_2;\omega_1)\right|^2\Gamma_p$$
$$\times\left[(\omega_{pn} - \omega_1 - \omega_2)^2 + \Gamma_p^2\right]^{-1}. \tag{3.217}$$

Note that even in the absence of real nonlinear absorption, when $N_p \ll N_n \approx N$, the above formula implies that κ is proportional to the intensity of the strong wave (in contrast to the linear absorption coefficient (Section 1.4), which is independent of the intensity.

3.7.5. The Relationship between the Nonlinear Absorption Coefficient and the Absorption Cross Section

If we divide the transition rate (3.204) of two-photon absorption by the flux of photons from the weak wave, that is, by $I(\omega_2)/\omega_2 = c|E_2|^2/2\pi\omega_2$, we find the absorption cross section for the weak electromagnetic wave with frequency ω_2:

$$\sigma(\omega_2) = \frac{4\pi\omega_2}{c} E_1^2 \Gamma_p \left| \chi_{np}^{(1)}(-\omega_2; \omega_1) \right|^2$$
$$\times \left[(\omega_{pn} - \omega_1 - \omega_2)^2 + \Gamma_p^2 \right]^{-1} \qquad (3.218)$$

By combining (3.217) with (3.218) we can easily find the relationship linking the nonlinear absorption coefficient κ for the wave with the frequency ω_2 and the absorption cross section σ:

$$\kappa = \sigma(N_n - N_p). \qquad (3.219)$$

Note that Eq. (3.219), as can be seen from the way in which it was derived, describes not only two-photon absorption but also multiphoton absorption and multiphoton ionization. In the case of K-photon absorption, in (3.219) we must substitute for σ the K-photon excitation cross section, whose general form is given in Section 2.4, while in the case of K-photon ionization we must substitute for σ the K-photon ionization cross section, and for N_p we must substitute the number of generated ions, N_i. Actual values of the multiphoton excitation and ionization cross sections are given in Basov, 1980, and Rapoport, Zon, and Manakov, 1978.

When the absorption is low (a low intensity of the wave with frequency ω_1), we have $N_p \ll N_n \approx N$, and (3.219) yields the simple formula

$$\kappa = N\sigma. \qquad (3.220)$$

From (3.219) and (3.220) it follows that in the case of linear absorption the absorption coefficient is directly proportional to the number of atoms per unit volume (Beer's law). Note that the same relationship links the linear absorption coefficient with the one-photon absorption cross section (Section 1.4).

If the field strength in the wave with frequency ω_1 is increased, state p becomes populated in the event of a two-photon resonance $\omega_1 + \omega_2 \approx \omega_{pn}$. According to the solution of the two-photon Rabi problem (see Section 3.2) in Delone and Kraĭnov, 1985),

$$N_p = \frac{N}{2} \frac{\left| \chi_{np}^{(1)}(-\omega_2; \omega_1) E_1 E_2 \right|^2}{(\omega_{pn} - \omega_1 - \omega_2)^2 + \left| \chi_{np}^{(1)}(-\omega_2; \omega_1) E_1 E_2 \right|^2}, \qquad (3.221)$$

and $N_n - N_p = N - 2N_p$, since $N_n + N_p = N$. Substituting (3.221) into (3.219) yields

$$\kappa = N\sigma \frac{(\omega_{pn} - \omega_1 - \omega_2)^2}{(\omega_{pn} - \omega_1 - \omega_2)^2 + \left|\chi_{np}^{(1)}(-\omega_2; \omega_1)E_1E_2\right|^2}. \qquad (3.222)$$

This expression can be made more precise if we employ the Breit-Wigner procedure and introduce the width Γ_p of state p:

$$\kappa = N\sigma \frac{(\omega_{pn} - \omega_1 - \omega_2)^2 + \Gamma_p^2}{(\omega_{pn} - \omega_1 - \omega_2)^2 + \Gamma_p^2 + \left|\chi_{np}^{(1)}(-\omega_2; \omega_1)E_1E_2\right|^2}. \qquad (3.223)$$

This makes κ finite even when the misfit of the two-photon resonance vanishes. If at a fixed value of $\omega_{pn} - \omega_1 - \omega_2$ we increase E_1, we must modify the expression (3.218) for σ in (3.222). According to the results obtained in Section 1.2.4, if

$$\left|\chi_{np}^{(1)}(-\omega_2; \omega_1)E_1E_2\right| > \Gamma_p, \qquad (3.224)$$

then in the denominator in (3.218) for Γ_p we must substitute a greater quantity, the *field width* $|\chi_{np}^{(1)}(-\omega_2; \omega_1)E_1E_2|$. The result is

$$\sigma(\omega_2) = \frac{4\pi\omega_2}{c} \frac{\Gamma_p\left|\chi_{np}^{(1)}(-\omega_2; \omega_1)E_1\right|^2}{(\omega_{pn} - \omega_1 - \omega_2)^2 + \Gamma_p^2 + \left|\chi_{np}^{(1)}(-\omega_2; \omega_1)E_1E_2\right|^2}. \qquad (3.225)$$

Substituting (3.225) into (3.223), we find that at a high intensity I_1 of the wave with frequency ω_1, when (3.224) is satisfied, κ decreases like I_1^{-1}.

3.7.6. Conclusion

Comparing the results obtained in this section with those obtained in Section 1.4 for linear absorption, we can conclude that nonlinear absorption, like linear, obeys an exponential law, but with a κ that depends on the intensity I_1 of the incident wave. This value of κ is proportional to the number of atoms per unit volume, N, so that Beer's law remains valid. However, Bouguer's law does not, since κ depends on I_1. This makes nonlinear absorption quite different from linear. The dependence of κ on I_1 is such that for small values of I_1 the absorption coefficient κ is a power function of I_1. On the other hand, in a strong field κ begins to diminish and tends to zero.

There are also two additional factors that make nonlinear absorption differ from linear absorption. First, nonlinear absorption may be caused both by multiphoton excitation of atoms and by multiphoton ionization of atoms, while with linear absorption the only mechanism is excitation. Second, if the intensity of a strong electromagnetic field propagating in a medium diminishes as z increases because of energy transfer to polarization waves with other frequencies, the exponential law of nonlinear absorption ceases to be valid.

3.8. CONCLUSION

The study of basic nonlinear-optics phenomena carried out in this chapter shows that in all respects these are qualitatively new phenomena with no analogy in linear optics. In nonlinear optics one observes higher harmonic excitation, stimulated Raman scattering and its higher harmonics, coupled waves, nonlinear refraction, rotation of the polarization plane, and an absorption coefficient depending on the light intensity. In linear optics there is no excitation of higher harmonics, only spontaneous Raman scattering occurs, light beams propagate independently of each other, light propagates along a straight line, the polarization of light remains unchanged, and the light absorption coefficient is independent of the light intensity.

The common factor that causes these differences is the nonlinear polarization of the medium, which occurs when the intensity of the incident light is high. Microscopically speaking, nonlinear polarization means that multiphoton processes begin to play an important role. The different in the specific form of nonlinear polarization of a medium determines the specific nonlinear-optics phenomena that appear in the interaction of intense light with the medium. It is important that not only the first nonlinear terms in the expansion of the polarization in powers of the field strength are responsible for the concrete nonlinear phenomena, but often also higher-order terms. Resonance effects make it possible to separate them.

To generate nonlinear-optics phenomena on the macroscopic level, wave coupling becomes important, including the coupling with nonlinear polarization waves. To excite nonlinear polarization waves, the exciting radiation must be coherent, in addition to being of high intensity. All this requires laser radiation.

Conclusion

We are coming to the end of our study of the basics of nonlinear optics of atomic gases. The natural question then is: what degree of generality do the statements have concerning atomic gases and optically transparent media in general? This question has been discussed in various ways throughout the book, but we believe it is important to summarize the situation. Of course, we will be interested solely in the qualitative aspects of the problem and ignore the quantitative ones. If from this standpoint we look at the entire collection of nonlinear-optics phenomena that appear in atomic gases, molecular gases, plasmas, liquids, and solids (crystals and glasses) under the action of coherent radiation ranging in wavelengths from the infrared to the ultraviolet and in pulse length from picoseconds to the continuous mode, then two broad statements can be made. First, with optimal choice of the parameters that characterize the medium and the radiation one can always obtain the basic nonlinear-optics phenomena described above, namely, higher-harmonic excitation, stimulated Raman scattering, coupling of waves with different frequencies, nonlinear refraction, nonlinear absorption, and polarization effects that accompany the propagation of a strong wave in a medium. Qualitatively all these phenomena are the same; they differ only in quantitative characteristics. This proves that our discussion has considerable generality. Second, in media that differ from rarefied atomic gases and with radiation that differs from the stationary case, additional nonlinear-optics phenomena, which have no analog in the previously discussed case, may occur. Let us elaborate further on this.

Let us start with rarefied atomic gases and briefly examine the phenomena that emerge when one steps outside the realm of the above model, steady-state monochromatic radiation interacting with a medium within the unperturbed-field approximation with respect to the incident wave. Outside the framework of this approximation, that is, allowing for attenuation of the incident wave as the wave propagates through the medium, only the quantitative characteristics of the described phenomena change (Akhmanov and Koroteev, 1981; Butylkin, Kaplan, Khronopulo, and Yakubovich, 1984). The same is true when one takes real quasimonochromatic radiation (instead of the model of monochromatic radiation), at least within the limits of the spectral width typical of laser radiation. Rapid fluctuations in the intensity of multimode laser radiation in a number of cases lead to results that differ from those obtained within the above model (e.g., in the resonance Stark effect), but these differences do not drastically change the main nonlinear-optics phenomena (Akhmanov and Chirkin, 1971, Akhmanov, D'yakov, and Chirkin, 1981, and Delone, Kovarskiĭ, Masalov, and Perel'man, 1980).

The situation changes considerably when we ignore the assumption that the incident electromagnetic wave is steady-state. As soon as we assume that the duration of the radiation can assume any value, we are forced to take into account a new characteristic, the relaxation time of the medium. In the limiting case, where the pulse of the exciting radiation is much shorter than the relaxation time, new phenomena may emerge. For instance, when the atoms of the medium are resonantly excited, the atoms on the wavefront may transfer from the ground state to an excited state while the rest of the pulse is not absorbed by the medium. This phenomenon is known as self-induced transparency (Allen and Eberly, 1975). When a second pulse that has a certain length and lags behind the first by a certain interval passes through such a medium, the excited atoms will coherently relax, and a light echo appears (Macomber, 1976). These non-steady-state nonlinear-optics phenomena have no analog when the interaction of field and medium is steady-state; they are widely used in the nonlinear spectroscopy of gases (Akhmanov and Koroteev, 1981; Brewer, 1977; Shoemaker, 1978).

The nonlinearity of the medium created by the radiation pulse may influence the radiation, changing its spectral and temporal characteristics (Kryukov and Letokhov, 1970).

Now let us turn to other transparent media than atomic gases.

Qualitatively, all nonlinear-optics phenomena in rarefied molecular gases are the same as in atomic gases. The more complicated spectrum of natural frequencies of molecules and the presence of a dipole moment in molecules in the absence of an external field lead to the well-known differences in the microscopic picture of the interaction. For instance, the Kerr effect plays a much more important role for molecules than for atoms (Kielich, 1981). However, this only leads to a change in some quantitative characteristics, of which we have already spoken.

To finish the topic of gases, we note that the density of the gas is limited from above by the possibility of optical breakdown (Raĭzer, 1977). The breakdown threshold is determined by the product of three parameters characterizing the radiation (intensity and pulse length) and the gas (density) (Delone and Kraĭnov, 1985). The type of gas and the spatial distribution of the laser radiation also play a certain role here, but in the first approximation these effects can be ignored. We must also bear in mind that as we move closer to the breakdown threshold, there is the possibility of self-focusing of the radiation, since the intensity of the radiation greatly increases in self-focusing.

Some remarks are in order concerning nonlinear-optics effects in plasmas. At the base of such phenomena lies the possibility of a change in the free-electron permittivity of a plasma (Lifshitz and Pitaevskiĭ, 1981). Nonlinear phenomena occurring in a plasma as a result of the action of strong light (Hora, 1975; Hughes, 1975; Ready, 1971) are of great practical importance, for instance, in the problem of controlled thermonuclear fusion (Bruekner and Jorna, 1974).

If we turn to condensed transparent media, the differences from gases arise at all levels of consideration. Here we will mention only the important differences. Looking for the microscopic reasons for nonlinear polarization in condensed media, we must first bear in mind the electrostriction phenomenon (Born and Wolf, 1975), which to a great extent determines the refractive index of a condensed medium in a strong field. The macroscopic anisotropy of crystals also plays an important role, since it makes possible the absorption of an even number of photons in a single process, for instance, the generation of the second harmonic of the incident radiation. Finally, in a solid there exists a special channel of absorption of the incident wave, the generation of elastic lattice vibrations (or the excitation of phonons, if one uses the terminology of the quantum theory). It is this channel that leads to a qualitatively new nonlinear phenomenon without any analog in gases: stimulated Brillouin scattering (Fabelinskiĭ, 1968; Yariv, 1975). Hypersound waves generated in transparent insulators by high-power laser radiation (Tucker and Rampton, 1972) may be so strong as to destroy glasses and crystals. The above-discussed nonlinear-optics phenomena also possess special features here, but they are only quantitative. As an example we can point to the various ways in which stimulated Raman scattering can be realized, of which we spoke in Section 3.3.

Finally, the reader should always bear in mind that the nonlinear phenomena studied in this book with the help of optics, that is, the science of visible light, may have analogs in other frequency ranges. Important similarities can be noticed between nonlinear optics and nonlinear acoustics (Rudenko and Soluyan, 1977). The latter studies the propagation of sound waves in a medium with allowance for the nonlinear terms in the equations of hydrodynamics and the equation of state. The general theory of waves (Whitham, 1974) provides a good framework for uniting such phenomena.

References

Adonts, G. G., and L. M. Kocharyan, *Opt. Spectrosc. (USSR)*, **40**, 412 (1976).

Agarwal, G., and S. Tewari, *Phys. Rev. A*, **29**, 1922 (1983).

Akhmanov, S. A., *Sov. Radiophys.*, **17**, 325 (1974).

Akhmanov, S. A., and A. S. Chirkin, *Statistical Phenomena in Nonlinear Optics*, Moscow: Moscow Univ. Press, 1971 (in Russian).

Akhmanov, S. A., K. N. Drabovich, A. P. Sukhorukov, and A. S. Chirkin, *Sov. Phys. JETP*, **59**, 485 (1970).

Akhmanov, S. A., Yu. E. D'yakov, and A. S. Chirkin, *Introduction to Statistical Radiophysics and Optics*, Moscow: Nauka, 1981 (in Russian).

Akhmanov, S. A., and R. V. Khokhlov, *Problems in Nonlinear Optics*, New York, 1964.

Akhmanov, S. A., and R. V. Khokhlov, *Sov. Phys. Usp.*, **9**, 210 (1966).

Akhmanov, S. A., R. V. Khokhlov, and A. P. Sukhorukov, in: *Laser Handbook*, ed. by F. T. Arecchi and E. O. Schulz-Dubois, Vol. 2, Amsterdam: North-Holland, 1972: p. 1153.

Akhmanov, S. A., and N. I. Koroteev, *Methods of Nonlinear Optics in the Spectroscopy of Scattered Light*, Moscow: Nauka, 1981 (in Russian).

Akhmanov, S. A., A. P. Sukhorukov, and R. V. Khokhlov, *Sov. Phys. Usp.*, **10**, 609 (1968).

Alber, G., and P. Zoller, *Phys. Rev. A*, **27**, 1373 (1983).

Allen, L., and J. H. Eberly, *Optical Resonance and Two-Level Atoms*, New York: Wiley, 1975.

Anikin, V. I., S. V. Kryuchkov, and V. E. Ogluzdin, *Sov. J. Quantum Electron.*, **4**, 1065 (1975).

Apanasevich, P. A., *The Fundamentals of the Theory of Interaction of Light with Matter*, Minsk: Nauka i Tekhnika, 1977 (in Russian).

Armstrong, J. A., N. Bloembergen, J. Ducing, and P. S. Pershan, *Phys. Rev.* **127**, 1918 (1962).

Armstrong, J. A., and J. J. Wynne, in: *Nonlinear Spectroscopy (Proc. International School of Physics "Enrico Fermi," Course LXIV, Varenna, 30 June–12 July, 1975)*, ed. by N. Bloembergen, Amsterdam: North-Holland, 1977: pp. 152–169.

Arutyunyan, V. M., and G. G. Adonts, *Opt. Spectrosc. (USSR)*, **46**, 809 (1979).

Arutyunyan, V. M., E. G. Kanetsyan, and V. O. Chaltykyan, *Sov. Phys. JETP*, **62**, 908 (1972).

Arutyunyan, V. M., T. A. Papazyan, G. G. Adonts, A. V. Karmanyan, S. P. Ishkhanyan, and L. Khol'ts, *Sov. Phys. JETP*, **41**, 22 (1975).

Averbakh, V. S., A. A. Betin, V. A. Gaponov, A. I. Makarov, and G. A. Pasmanik, *Sov. Radiophys.*, **21**, 1077 (1978).

Bakhramov, S. A., U. G. Gulyamov, K. N. Drabovich, and Ya. Z. Faĭzulaev, *JETP Lett.*, **21**, 102 (1975).

Basov, N. G. (ed.), *Multiphoton Ionization of Atoms (Vol. 115 of the Proceedings of the Lebedev Physics Institute)*, Moscow: Nauka, 1980 (in Russian).

Berestetskiĭ, V. B., E. M. Lifshitz, and L. P. Pitaevskiĭ, *Quantum Electrodynamics*, 2nd ed., Oxford: Pergamon Press, 1980.

Bertein, F., *Bases de l'electronique quantique. Absorption et émission de champ electromagnétique*, Paris: Eyrolles, 1969.

Besplavo, V. I., and V. I. Talanov, *JETP Lett.*, **3**, 471 (1966).

Beterov, I. M., N. V. Fateev, and V. P. Chebotaev, *JETP Lett.*, **36**, 307 (1982).

Bethe, H. A., and E. A. Salpeter, *Quantum Mechanics of One- and Two-Electron Atoms*, Berlin: Springer, 1957.

Bethune, D., J. Lankard, and P. Sorokin, *Opt. Lett.*, **4**, 103 (1979).

Bethune, D., R. Smith, and Y. Shen, *Phys. Rev. A*, **17**, 277 (1978).

Bjorkholm, J., and A. Ashkin, *Phys. Rev. Lett.*, **32**, 129 (1974).

Bjorklund, G., *IEEE J. Quantum Electron.*, **QE-11**, 287 (1975).

Bloembergen, N., *Nonlinear Optics*, New York: W. A. Benjamin, 1965.

Bloembergen, N. (ed.), *Nonlinear Spectroscopy (Proc. International School of Physics "Enrico Fermi," Course LXIV, Varenna, 30 June–12 July, 1975)*, Amsterdam: North-Holland, 1977.

Bloembergen, N., G. Breit, P. Lallemand, A. Pine, and P. Simova, *IEEE J. Quantum Electron.*, **QE-3**, 197 (1967).

Bloom, D., G. Bekkers, T. Yong, and S. Harris, *Appl. Phys. Lett.*, **26**, 687 (1975).

Bloom, D. M., J. T. Yardley, J. F. Young, and S. E. Harris, *Appl. Phys. Lett.*, **24**, 427 (1974).

Bobovich, Ya. S., and A. V. Bortkevich, *Sov. Phys. Usp.*, **14**, 1 (1971).

Born, M., and E. Wolf, *Principles of Optics*, 5th ed., Oxford: Pergamon Press, 1975.

Brewer, R. G., in: *Nonlinear Spectroscopy (Proc. International School of Physics "Enrico Fermi," Course LXIV, Varenna, 30 June–12 July, 1975)*, ed. by N. Bloembergen, Amsterdam: North-Holland, 1977: pp. 87–137.

Brueckner, K. A., and S. Jorna, *Rev. Mod. Phys.*, **46**, 325 (1974).

Butylkin, V. S., A. E. Kaplan, and Yu. G. Khronopulo, *Sov. Phys. JETP*, **32**, 501 (1971).

Butylkin, V. S., A. E. Kaplan, Yu. G. Khronopulo, and E. I. Yakubovich, *Resonant Nonlinear Interactions of Light and Matter*, Berlin: Springer, 1984.

Ciao, R., E. Garmiere, and C. Townes, *Phys. Rev. Lett.*, **13**, 479 (1964).

Crance, M., and L. Armstrong, *J. Phys. B*, **15**, 4637 (1982).

Cummings, H., N. Knable, and Y. Yeh, *Phys. Rev. Lett.*, **12**, 150 (1964).

Davydkin, V. A., B. A. Zon, N. L. Manakov, and L. P. Rapoport, *Sov. Phys. JETP*, **33**, 70 (1971).

Delone, N. B., *Izv. Akad. Nauk SSSR, Ser. Fiz.*, **49**, 471 (1985).

Delone, N. B., V. A. Kovarskiĭ, A. V. Masalov, and N. F. Perel'man, *Sov. Phys. Usp.*, **23**, 472 (1980).

Delone, N. B., and V. P. Kraĭnov, *Atoms in Strong Light Fields*, Berlin: Springer, 1985.

Dimov, S. S., D. I. Metchkov, L. I. Pavlov, and K. V. Stamenov, *Sov. J. Quantum Electron.*, **12**, 672 (1982).

Dimov, S. S., L. I. Pavlov, Yu. I. Geller, and A. K. Popov, *Kvant. Electron. (Moscow)*, **10**, 1635 (1983).

Ditchburn, R., *Light*, London: Blackie, 1963.

Dmitriev, V. G., and L. V. Tarasov, *Applied Nonlinear Optics*, Moscow: Radio i Svyaz', 1982 (in Russian).

Doitcheva, M., V. Mitev, L. Pavlov, and K. Stamenov, *Opt. Quantum Electron.*, **10**, 131 (1978).

Drabovich, K. N., M. Ignatovichyus, R. Kupris, A. Matsulyavichyus, S. M. Pershin, N. M. Synyavskiĭ, V. Smil'gyavichyus, and A. L. Surovegin, *Bull. Acad. Sci. USSR*, *Phys. Ser.*, **46**, No. 8, 180 (1982).

Dunn, M. H., *Opt. Commun.*, **45**, 346 (1983).

Eicher, H., *IEEE J. Quantum Electron.*, **QE-11**, 121 (1975).

Fabelinskiĭ, I. L., *Molecular Scattering of Light*, New York: Plenum Press, 1968.

Finn, R., and T. Ward, *Phys. Rev. A*, **2**, 285 (1971).

Gavrila, M., *Phys. Rev.*, **163**, 147 (1967).

Geller, Yu. I., and A. K. Popov, *Laser Inducing of Nonlinear Resonances in Continuum*, Novosibirsk: Nauka, 1981 (in Russian).

Grasyuk, A. Z., *Kvant. Electron. (Moscow)*, **1**, 485 (1974).

Grischkowsky, D., *Phys. Rev. Lett.*, **24**, 866 (1970).

Grischkowsky, D., and J. Armstrong, *Phys. Rev. A*, **6**, 1566 (1972).

Hanna, D. S., M. A. Yuratich, and D. Cotter, *Nonlinear Optics of Free Atoms and Molecules*, Berlin: Springer, 1979.

Hartig, W., *Appl. Phys.*, **15**, 427 (1978).

Heitler, W., *The Quantum Theory of Radiation*, 3rd ed., Oxford: Clarendon Press, 1954.

Hora, H., *Laser Plasmas and Nuclear Energy*, New York: Plenum Press, 1975.

Hughes, T., *Plasmas and Laser Light*, London: Higler, 1975.

Il'inskiĭ, Yu. A., and V. D. Taranukhin, *Sov. J. Quantum Electron.*, **5**, 805 (1976).

Javan, A., and P. Kelley, *IEEE J. Quantum Electron.*, **QE-2**, 470 (1966).

Jong, T., G. Bjorklund, A. Kung, R. Miles, and S. Harris, *Phys. Rev. Lett.*, **27**, 1551 (1971).

Kaiser, W., and M. Maier, in: *Laser Handbook*, ed. by F. T. Arecchi and E. O. Schulz-Dubois, Vol. 2, Amsterdam: North-Holland, 1972: p. 1077.

Kastler, A., *Proc. Phys. Soc. London*, **67A**, 858 (1954).

Keldysh, L. V., *Sov. Phys. JETP*, **20**, 1307 (1965).

Kelley, P., *Phys. Rev. Lett.*, **15**, 1005 (1965).

Kielich, St., *Molekularna optyka nieliniowa*, Warsaw: Panstwowe wydawnictwo naukowe, 1981.

Kildal, H., and S. Brueck, *IEEE J. Quantum Electron.*, **QE-16**, 566 (1980).

Klarsfeld, S., and A. Maquet, *Phys. Rev. Lett.*, **29**, 79 (1972).

Korolev, F. A., V. A. Mikhailov, and V. I. Odintsov, *Opt. Spectrosc. (USSR)*, **44**, 535 (1978).

Kovarskiĭ, V. A., *Multiquantum Transitions*, Kishinev: Shtiintsa, 1974 (in Russian).

Kraĭnov, V. P., *Sov. Phys. JETP*, **43**, 622 (1977).

Kryukov, P. G., and V. S. Letokhov, *Sov. Phys. Usp.*, **12**, 641 (1970).

Kung, A., J. Jong, and S. Harris, *Appl. Phys. Lett.*, **22**, 301 (1973).

Landau, L. D., and E. M. Lifshitz, *The Classical Theory of Fields*, 4th ed., Oxford: Pergamon Press, 1975.

Landau, L. D., and E. M. Lifshitz, *Quantum Mechanics: Nonrelativistic Theory*, 3rd ed., Oxford: Pergamon Press, 1977.

Le Boitex, S., R. Raj, Q. Gao, D. Bloch, and M. Ducloy, *J. Opt. Soc. Am. B*, **1**, 501 (1984).

Letokhov, V. S., *Nonlinear Laser Chemistry*, Berlin: Springer, 1982.

Letokhov, V. S., and V. P. Chebotaev, *Nonlinear Laser Spectroscopy*, Berlin: Springer, 1977.

Liao, P., and G. Bjorklund, *Phys. Rev. A*, **15**, 2009 (1977).

Lifshitz, E. M., and L. P. Pitaevskiĭ, *Physical Kinetics*, Oxford: Pergamon Press, 1981.

Litvak, A. G., *JETP Lett.*, **4**, 230 (1966).

Loudon, R., *The Quantum Theory of Light*, Oxford: Clarendon Press, 1973.

Louisell, W. H., *Coupled Mode and Parametric Electronics*, New York: Wiley, 1960.

Lugovoĭ, V. N., *Introduction to the Theory of Stimulated Raman Scattering*, Moscow: Nauka, 1968 (in Russian).

Lugovoĭ, V. N., and A. M. Prokhorov, *Sov. Phys. Usp.*, **16**, 658 (1974).

Lumpkin, O., *IEEE J. Quantum Electron.*, **QE-4**, 226 (1978).

Macomber, J. D., *The Dynamics of Spectroscopic Transitions*, New York: Wiley, 1976.

Manakov, N. L., and V. D. Ovsyannikov, *Opt. Spectrosc. (USSR)*, **48**, 359 (1980).

Manakov, N. L., V. G. Ovsyannikov, and L. P. Rapoport, *Sov. J. Quantum Electron.*, **5**, 22 (1975).

Manley, T., and H. Rowe, *Proc. IRE*, **47**, 2115 (1959).

Marburger, J., *Proc. Quantum Electron.*, **4**, 35 (1975).

Marcano, A. O., and F. Garcia-Golding, *J. Opt. Soc. Am.*, **72**, 957 (1982).

Martin, R. M., and L. M. Falicov, in: *Light Scattering in Solids*, ed. by M. Cardona, Berlin: Springer, 1975: pp. 79–145.

Mash, D. I., V. V. Morozov, V. S. Starunov, and I. L. Fabelinskiĭ, *JETP Lett.*, **2**, 25 (1965).

Mavroyannis, C., *Phys. Rev. A*, **27**, 1414 (1983).

Miles, R., and S. Harris, *IEEE J. Quantum Electron.*, **QE-9**, 470 (1973).

Miyazaki, K., and H. Kashiwagi, *Phys. Rev. A*, **18**, 635 (1978).

Miyazaki, K., T. Sato, and H. Kashiwagi, *Phys. Rev. Lett.*, **43**, 1154 (1979).

Mossberg, T., A. Flusberg, and S. Hartmann, *Opt. Commun.*, **25**, 121 (1978).

Morellec, J., D. Normand, and G. Petite, *Adv. At. Mol. Phys.*, **18**, 97 (1982).

Muradyan, A. Zh., G. G. Adonts, and V. G. Kolomiets, *Izv. Akad. Nauk Arm. SSR, Ser. Fiz.*, **8**, 331 (1973).

Normand, D., J. Morellec, and J. Reif, *J. Phys. B*, **16**, L227 (1983).

Ohashi, Y., Y. Ishibashi, T. Kobayasi, and H. Inaba, *Jpn. J. Appl. Phys.*, **15**, 1817 (1976).

Orr, B. I., and J. E. Ward, *Mol. Phys.*, **20**, 513 (1971).

Pavlov, L., S. Dimov, D. I. Metchkov, G. M. Mileva, and K. V. Stamenov, *Phys. Lett. A*, **89**, 441 (1982).

Phillion, D. W., D. L. Banner, E. M. Campbell, R. E. Turner, and K. G. Estabrook, *Phys. Fluids*, **25**, 1434 (1982).

Puell, H., H. Scheingraber, and C. Vidal, *Phys. Rev. A*, **22**, 1165 (1980).

Puell, H., K. Spanner, W. Falkenstein, W. Kaiser, and C. Vidal, *Phys. Rev. A*, **14**, 2240 (1976).

Raĭzer, Yu. P., *JETP Lett.*, **4**, 85 (1966).

Raĭzer, Yu. P., *Sov. Phys. JETP*, **25**, 308 (1967).

Raĭzer, Yu. P., *Laser-Induced Discharge Phenomena*, New York: Plenum Press, 1977.

Rapoport, L. P., B. A. Zon, and N. L. Manakov, *Theory of Multiphoton Processes in Atoms*, Moscow: Atomizdat, 1978 (in Russian).

Rautian, S. G., in: *Nonlinear Optics (Vol. 43 of the Proceedings of the Lebedev Physics Institute)*, ed. by D. V. Skobel'tsyn, New York: Plenum Press, 1970: p. 1.

Rautian, S. G., in: *Invited papers of the II Conference on Interaction of Electrons with Strong Electromagnetic Fields*, Budapest: Central Research Institute for Physics, 1975: p. 150.

Rautian, S. G., G. I. Smirnov, and A. M. Shalagin, *Nonlinear Resonances in Spectra of Atoms and Molecules*, Novosibirsk: Nauka, 1979 (in Russian).

Ready, J. F., *Effects of High-Power Laser Radiation*, New York: Academic Press, 1971.

Reif, J., and H. Walther, *Appl. Phys.*, **15**, 361 (1978).

Reintjes, J., C. She, and R. Eckardt, *IEEE J. Quantum Electron.*, **QE-14**, 581 (1978).

Rudenko, O. W., and S. I. Soluyan, *Theoretical Foundations of Nonlinear Acoustics*, New York: Plenum Press, 1977.

Schubert, M., and B. Wilhelmi, *Einführung in die nichtlineare Optik, Vol. 1: Klassische Beschreibung*, Leipzig: Teubner, 1971.

Shalagin, A. M., *Sov. Phys. JETP*, **46**, 50 (1977).

Shen, Y. P., in: *Light Scattering in Solids*, ed. by M. Cardona, Berlin: Springer, 1975a: pp. 275–328.

Shen, Y. P., *Proc. Quantum Electron.*, **4**, 1 (1975b).

Shen, Y. P., in: *Nonlinear Spectroscopy (Proc. International School of Physics "Enrico Fermi," Course LXIV, Varenna, 30 June–12 July, 1975)*, ed. by N. Bloembergen, Amsterdam: North-Holland, 1977: pp. 170–200.

Shoemaker, R., in: *Laser and Coherence Spectroscopy*, ed. by J. Steinfeld, New York: Plenum Press, 1978: p. 197.

Skupsky, S., and R. Osborn, *Nuovo Cimento B*, **26**, 181 (1975).

Smith, D., *IEEE J. Quantum Electron.*, **QE-5**, 600 (1969).

Sobelman, I. I., *Atomic Spectra and Radiative Transitions*, Berlin: Springer, 1979.

Sobelman, I. I., L. A. Vainstein, and E. A. Yukov, *Excitation of Atoms and Broadening of Spectral Lines*, Berlin: Springer, 1981.

Starunov, V. S., and I. L. Fabelinskiĭ, *Sov. Phys. Usp.*, **12**, 463 (1970).

Tewari, S., *J. Phys. B*, **16**, L785 (1983).

Tomov, T., and M. Richardson, *IEEE J. Quantum Electron.*, **QE-12**, 521 (1976).

Tucker, J. W., and V. M. Rampton, *Microwave Ultrasonics in Solid State Physics*, Amsterdam: North-Holland, 1972.

Vasilenko, L. S., V. P. Chebotaev, and A. V. Shishaev, *JETP Lett.*, **12**, 113 (1970).

Vidal, C., and T. Cooper, *J. Appl. Phys.*, **40**, 3370 (1969).

Vol'kenshtein, M. V., *Molecular Optics*, Moscow: GITTL, 1951 (in Russian).

Walther, H. (ed.), *Laser Spectroscopy of Atoms and Molecules*, Berlin: Springer, 1976.

Ward, J., and G. New, *Phys. Rev.*, **185**, 57 (1969).

Whitham, G., *Linear and Nonlinear Waves*, New York: Wiley, 1974.

Wieman, G., and T. Hänsch, *Phys. Rev. Lett.*, **36**, 1170 (1976).

Yakovlenko, S. I., *Radiation-Collision Phenomena*, Moscow: Energoatomizdat, 1984 (in Russian).

Yariv, A., *Quantum Electronics*, 2nd ed., New York: Wiley, 1975.

Zeldovich, B. Ya., and I. I. Sobelman, *Sov. Phys. Usp.*, **13**, 307 (1970).

Zernike, F., and J. E. Midwinter, *Applied Nonlinear Optics*, New York: Wiley, 1973.

Zon, B. A., N. L. Manakov, and L. P. Rapoport, *Sov. Phys. JETP*, **34**, 515 (1972).

Author Index

213

Subject Index